PRAISE FOR **MURRAY LONGE**

"Murray distils decades of experience and complex concepts into an accessible roadmap."

KERI A. CLARKE - Start-up industry professional

"Murray has a deep understanding of what he is discussing."

KROL- Amazon Reviewer

PRAISE FOR **MURRAY LONGE'S OTHER BOOKS**

"Like sitting down for an in-depth chat with a knowledgeable, experienced friend in the business. Practical advice based on real world practical success. The writing style makes it easy to read and absorb what might otherwise be quite complex ideas."

WORDSMITH - Amazon Reviewer

"A concise and practical guide for entrepreneurs, packed with clear insights, case studies, and actionable steps. An invaluable resource."

KERI A. CLARKE

"This book is a very well-written, thorough look at every aspect of starting a business. I highly recommend it, especially as it's written by a successful entrepreneur, someone who has actually been through the start-up process from start to finish."

PETER G. - Amazon Reviewer

The Startup Success Series Book 3

Scale Without Breaking

How to Grow Your Startup Without Losing What Made It Work

Murray Longe

WHITMAN CROSS

Published by Whitman Cross Publishing, Melbourne, Australia.

First edition 2025
Second edition 2026

ISBN 978-1-7645438-0-4 (ebook)
ISBN 978-1-7645438-1-1 (paperback)

NATIONAL LIBRARY OF AUSTRALIA

A catalogue record for this book is available from the National Library of Australia

Disclaimer

The information in this book represents the author's views as of the date of publication. It is for general informational purposes only and is not intended as professional, legal, financial, or business advice. Because circumstances and regulations change, the author reserves the right to revise his views in light of new conditions. Readers should consider their own personal circumstances and consult with qualified professionals before taking any action based on the information in this book. While every effort has been made to ensure the accuracy and completeness of the information, the author and publisher assume no responsibility for any errors, omissions, or inaccuracies. Use of the information is at the reader's own risk.

For more information, visit whitmancross.com

WHITMAN CROSS

Also by Murray Longe

How To Start Your Start-Up:
The 10 Key Steps To Get From Idea To Investment

Project Management for Startups:
Simple Tools and Templates to Launch, Lead, and Thrive

The Startup Success Series

#1 *Find Problems Worth Solving:*
How to Validate Your Startup Idea

#2 *Beyond The Minimum Viable Product:*
Build Products Customers Really Want

#3 *Scale Without Breaking:*
How to Grow Your Startup Without Losing What Made It Work

#4 *Beyond The Pitch Deck:*
How to Raise Startup Capital

For every founder staring at the gap between where you are and where you want to be.

You know your business could be bigger. You're just not sure if you're ready, or if trying will break everything you've built.

Read this book. Then make the call.

Murray Longe
December 2025

Table of Contents

Table of Contents

Table of Contents

Table of Contents

Introduction: The Second Build

You made it.

Maybe you survived the startup phase and found customers who actually pay for what you built. Maybe you've been running a steady business for 18 months or three years, and you're finally comfortable enough to think bigger. Maybe you just acquired a company and you're ready to grow it. Whatever path brought you here, you've proven something works.

Now you're looking at the next level. You see bigger opportunities. Larger markets. More revenue. Real scale.

Congratulations. Now comes the hard part.

Scaling a business feels nothing like maintaining one. Running steady is about consistency and control. Scaling is about making everything work at volume without breaking what you built. Whether you're two years in or twenty, the transition from stable to scaling changes everything.

The skills that got you here - personal relationships with customers, hands-on involvement, knowing every detail - those same skills can kill you at scale.

I learned this the expensive way.

My first business sold online directory advertising to small and medium businesses through a call center. We started with four sales agents. They were good. Really good. We were hitting our numbers, making sales, feeling unstoppable. We had been running successfully for a month. We were stable. We had revenue coming in, so we did what seemed obvious: we grew the team to scale the revenue.

Within months, we went from four agents to 60. On paper, it looked like progress. In reality, we were falling apart. Our revenue increased, but our conversion rates per agent dropped hard. Our costs had risen fast with the extra people and the revenue couldn't keep up. What we missed was the cash. We were burning through our runway so fast we barely noticed until we were almost out of road.

We had to let 36 people go just to survive. Then we rebuilt slowly, this time paying attention to the numbers that actually mattered.

That failure taught me more about scaling than any success ever could.

The Fear Every Founder Feels

Here's what nobody says out loud: you're terrified that growth will destroy what makes your business special.

Right now, you know your customers. You understand the product. You can feel when something's off. You might even still be involved in delivery or sales. That connection is what made you successful.

Scale threatens all of it.

At 50 people, you can't know everyone. At 100, you can't touch every customer. At 500, you're running a machine you barely recognize. The question that keeps you up at night is this: Can you grow without losing the magic that got you here?

The answer is yes. But only if you do it right.

Most founders don't. They scale too fast on the wrong metrics. They hire too many people before building the systems to support them. They chase revenue while bleeding cash. They delegate too late or too early. They build infrastructure that creates bureaucracy instead of leverage.

The result? Companies that grow themselves to death.

Whether you run a tech startup, a manufacturing business, a service company, or a retail operation, the core challenge is the same. The specific metrics you watch and the systems you build will differ by industry, but the discipline of scaling without breaking applies to all of them. We'll explore those differences where they matter most.

Why Most Scaling Attempts Fail

The statistics are brutal. Research suggests that only about 10% of startups ever reach meaningful scale. Of those that do, many stumble badly along the way - layoffs, pivots, near-death experiences that could have been avoided.

The failures follow patterns:

Some companies scale before they're ready. They've got early traction but no proven model. They raise money and grow headcount, mistaking activity for progress. The metrics look impressive until they don't. Then the money runs out.

Some companies scale on vanity metrics. User growth looks great. Revenue climbs. The board is happy. But unit economics are broken. Every new customer loses money. Growth accelerates the losses.

Some companies scale without systems. They hire people faster than they can train them. Quality drops. Culture dilutes. The founder becomes a bottleneck on everything. Nothing works without their direct involvement.

Some companies scale up and lose their identity. They professionalize too hard. Process takes over judgment. The scrappy team that won early customers turns into a corporate machine that can't move fast anymore. Customers notice. Competitors pounce and seize the opportunity.

All of these are avoidable. But you have to know what to watch for.

Two Paths to Scale

This book covers both ways to scale: building it yourself and buying your way there.

Most startup books assume you're building from scratch. You've got your product, your team, your runway. Now you need to grow it. That's the organic path, and we'll cover it thoroughly.

But there's another way: acquisition entrepreneurship.

Instead of scaling your startup, you buy a business that's already established. Or you buy a competitor and combine operations. Or you buy companies to broaden your capabilities faster than you could build them yourself.

This isn't just for late-stage companies with piles of cash. Small acquisitions can accelerate growth at any stage. The baby boomers are retiring. Millions of profitable businesses need buyers. Capital has never been more available. Business loans, seller financing, and earnouts make deals possible that seemed impossible a decade ago.

We'll explore when to build, when to buy, and how to integrate what you acquire without destroying it.

What This Book Will Do

This is not a textbook. It's not a comprehensive guide to every aspect of scaling. It's a field manual for founders who want to grow without blowing up.

You'll learn how to know if you're ready to scale. How to identify the metrics that actually matter. How to build systems without creating bureaucracy. How to grow your team without losing your culture. How to fund growth without losing control. How to recognize when scaling isn't working and what to do about it.

Every chapter includes real examples - companies that scaled successfully and companies that didn't. You'll see what worked, what failed, and why. You'll get frameworks you can use immediately, not theories you have to adapt.

Most importantly, you'll get honesty. Scaling is hard. Sometimes it doesn't work. Sometimes the right answer is to pivot, pause, or even sell. We'll talk about that too.

What This Book Won't Do

This book won't give you a guaranteed formula for success. Scaling isn't a simple step-by-step process. Your market is different. Your business model is different. Your team is unique. What worked for Airbnb or Stripe or some other Silicon Valley darling might not work for you.

This book won't promise easy growth. If you're looking for hacks, shortcuts, or secrets, you're in the wrong place. Scaling is execution. It's discipline. It's watching numbers and making hard decisions and sometimes admitting you got it wrong.

This book won't sugarcoat failure. Some companies shouldn't scale. Some founders aren't suited to lead scaled organizations. Sometimes the smartest move is to stay small, sell, or shut down. That's not defeat. That's wisdom.

Your Journey Starts Here

You've built something worth scaling. That's rare. Most people never get this far.

The next phase will test you differently than the startup phase did. You'll need different skills. Different disciplines. Different metrics. Different honesty with yourself about what's working and what isn't.

You'll make mistakes. I did. Every founder does. The question is whether you'll recognize them in time to fix them.

This book exists to help you do exactly that.

Let's get started.

1

Are You Ready To Scale?

The Honest Assessment

The worst time to realize you weren't ready to scale is six months into scaling.

You've hired 30 people. You've signed a bigger lease. You've committed to new systems and contracts. Your burn rate tripled. And somewhere in month four or five, you notice the numbers aren't working. Conversion rates are down. Quality is slipping. Your best people look exhausted. Customers are complaining more.

By then, retreat is expensive. You've built obligations you can't easily undo.

This is where most scaling attempts die. Not because the business was bad. Not because the market disappeared. But because the founder mistook desire for readiness.

I made this mistake. We had a successful call center operation with four agents selling directory advertising. We were making money. We felt confident. We saw the opportunity to grow revenue by adding more agents. So we did. We went from four to 60 in a few months.

What we didn't ask was whether we were ready for that growth. We had desire. We had ambition. We had customers. What we didn't have was the cash buffer to support 60 people.

The question isn't whether you want to scale. The question is whether your business can handle it right now.

Growth vs. Scaling: Why the Difference Matters

Most founders use these words like they mean the same thing. They don't.

Growth is adding revenue at the same pace you add resources. You hire five people, revenue goes up proportionally. You open a new location, sales increase from that location. Linear. Predictable. Sustainable.

Scaling is adding revenue faster than you add resources. You build systems that let you serve 10 times more customers with only twice the team. You create leverage. Your business model gets more efficient as it gets bigger.

A service business that bills by the hour grows. A software business that sells subscriptions scales. A restaurant that opens new locations grows. A franchise system that licenses the model scales.

Neither is better. They're just different. But if you're trying to scale a business that only knows how to grow, you'll burn cash and wear out your team chasing something your model can't deliver.

Here's the test: If you doubled your customer base tomorrow, would your costs double or would you handle most of it with your current team and systems?

If costs would double, you have a growth business. That's fine. Just don't fund it like a scaling business or you'll run out of money.

If you could handle the volume with minimal additional cost, you have a scaling business. Now the question is whether your systems and team can actually execute that scale.

Industry matters here. Software and digital products scale naturally. Manufacturing requires capital but can scale with automation. Professional services struggle to scale unless you productize or franchise. Retail scales through locations or online channels. Know what kind of business you're running.

The 5 Signs You're Ready to Scale

Readiness isn't about hitting a revenue number or raising a funding round. It's having a foundation that supports growth without breaking.

1. Your Unit Economics Work

You make more money from each customer than it costs to acquire and serve them. Not on average. Not if you squint at the numbers. Actually profitable per customer.

If you're losing money on every sale and planning to make it up in volume, you're not ready to scale. You're ready to go bankrupt faster.

Calculate your customer acquisition cost. Calculate your lifetime value. If LTV is at least three times CAC, you've got room to scale. If it's closer to one-to-one, you need to fix the model before you grow it.

This applies across industries. A manufacturing business needs healthy gross margins per unit. A service business needs profitable billable hours. A subscription business needs retention that justifies acquisition costs.

2. You Have Repeatable Processes

Growth can run on founder magic and hustle. Scaling can't.

If your sales process is "the founder sells," you're not ready. If your delivery process is "we figure it out each time," you're not ready. If your customer success process is "people email us and we respond," you're not ready.

Repeatable doesn't mean rigid. It means documented. It means someone other than you can do it. It means new hires can learn it in days, not months.

We had some of this in my directory business. Our four initial agents were self-starters, strong salespeople who figured things out. When we hired 56 more people, we expected them to do the same. They couldn't. We had scripts, we had clear processes, and a training program that was in its infancy. We expected every new agent we hired to be as good as our first group. But they weren't. They couldn't sell as well. Our hiring standards had dropped.

3. Your Team Can Grow Without You

If you're the bottleneck on every decision, you can't scale. You'll just create 60 people waiting for you instead of four.

Ready-to-scale businesses have managers who can hire, train, and run teams without constant founder oversight. They have decision-making frameworks that let people act without asking permission every time.

If you don't have at least one layer of management that can operate independently, you're not ready to scale.

This was our other failure. We hired 56 agents and made team leaders of those who looked to be the best. We could train people fast enough; we had the system for that. But we didn't monitor their performance properly. We didn't direct those managers on what to look for when managing their teams. We drowned.

4. You Have Cash Runway for 12-18 Months

Scaling costs money before it makes money. You hire ahead of revenue. You build systems before you need them. You invest in infrastructure that won't pay off for quarters.

If you're running on fumes, scaling will kill you. You need a buffer.

We thought we had runway. We watched our bank account. But we could model what 60 people would cost. We didn't account for the ramp time before new agents became productive. We didn't build in any margin for error. We were looking at the wrong metrics.

By the time we realized we were burning cash too fast, we had maybe eight weeks left. We had to fire 36 people just to survive. That's not scaling. That's panic.

5. Your Market Can Support the Scale You Want

This sounds obvious, but plenty of companies scale into markets that can't absorb them.

Do the math. If you're targeting small businesses in a specific region, how many are there? If you need 5,000 customers to hit your revenue target and there are only 3,000 potential customers in your market, you have a problem.

Scaling makes sense when your market is big enough to support your ambitions. If it's not, you need to expand your market or adjust your ambitions.

The 5 Signs You're Not Ready (And What Happens If You Ignore Them)

1. You're Still Searching for Product-Market Fit

If customers love what you do but you're still experimenting with pricing, positioning, or core features, you're not ready to scale. You'll scale confusion and waste money building the wrong thing bigger.

Wait until you know what to build. Then scale it.

2. Your Churn Rate Is High

If customers leave as fast as you acquire them, more marketing won't help. You'll just fill a leaky bucket faster.

Fix retention before you scale acquisition. Otherwise, you're paying twice - once to acquire customers and again to replace them when they leave.

This applies to employees too. If people quit within their first year, you have a culture or management problem. Scaling will make it worse.

3. Your Founder Is Still Doing Everything

If you're still the only salesperson, the only person who can close deals, and the only one who understands the product, you can't scale. You can grow a bit, but real scale requires systems that operate without you.

This is the hardest transition for founders. You built the business on your ability to do everything. Scaling requires you to build a business that works without you doing anything.

If you can't take a two-week vacation without the business falling apart, you're not ready to scale.

4. You Don't Have the Data You Need

If you can't tell me your customer acquisition cost, your lifetime value, your conversion rates at each stage of the funnel, your cash burn rate, and your unit economics, you're flying blind.

Scaling blind is suicide. You'll make decisions based on feelings instead of facts. You'll miss problems until they're catastrophic.

Get your data house in order before you scale up.

5. You're Scaling Because Investors Want You To

This is the trap that kills many funded companies. You raise money. Investors want growth. So you scale before you're ready because that's what they expect.

Bad idea.

Investors can find other investments. You only get one shot at this business. Don't let outside pressure push you to scale prematurely. Take the time to build the foundation.

The Buy-Then-Scale Alternative

Here's a different question: What if you didn't scale your business? What if you bought one that's already scaled?

Acquisition entrepreneurship bypasses many readiness questions. You're buying infrastructure, systems, customers, and a team. The business already runs. Your job is to improve it, not build it from scratch.

This path makes sense when:

- You have operational skills but don't want to build from zero
- You can see inefficiencies in existing businesses that you could fix
- You want cash flow from day one instead of years of losses
- The market has good businesses for sale (and right now, with baby boomers retiring, it does)

The readiness questions change.
Instead of "Can I scale what I built?" you ask, "Can I improve what I'm buying?" Instead of asking, "Do I have the systems?" you ask, "Can I integrate these systems with mine?"

We'll explore the acquisition strategy in more detail in Chapter 5, but the point here is simple: scaling isn't the only path to a bigger business.

The Timing Trap: Too Early vs. Too Late

Most founders err on one side or the other.

Scale too early and you outrun your ability to deliver. You hire people you can't train. You take customers you can't serve properly. You burn cash building infrastructure you're not ready to use. Quality drops. Culture breaks. Customers leave.

Scale too late and competitors take the market. You miss the window. Your team gets frustrated waiting. Momentum dies. The opportunity passes.

The right time to scale is when you've got the foundation in place and the market is still open. That window exists. But it doesn't stay open forever.

Here's how to know you're in it:

- Your metrics are solid and consistent over at least two quarters
- You have repeatable processes that new people can execute
- You have management depth to support more team members
- You have 12-18 months of cash runway
- Your market is growing and you're not losing share
- Customers are asking for more than you can currently deliver

If you're checking those boxes, you're probably ready. If you're missing three or more, wait. Fix the gaps first.

I've included a copy of The Scaling Readiness Scorecard on the next page. Take the time to go through each section to see if you're ready to scale now.

Case Study: Scaling Too Early

Webvan was an online grocery delivery startup that began operating in 1999. It was a big idea in a huge market, and investors poured about $800 million into it. They built giant automated warehouses, ordered a fleet of trucks, and expanded rapidly into multiple U.S. cities before ever proving the business worked in one market.

Instead of validating unit economics in the Bay Area, Webvan scaled execution on top of an unproven model. At one point they needed around 3,300 orders per day in their Bay Area operation to break even, but were only getting about 2,300. Their fulfillment and delivery costs were far higher than their order volume could support, customer growth lagged expectations, and people ordered less often than their models assumed. In 2001, after burning through its capital, Webvan shut down and filed for bankruptcy. It became a classic example of scaling too early and too fast on an unproven model.

Case Study: Scaling at the Right Time

Amazon launched its website in 1995 as an online bookstore. For the first few years, they concentrated on books, using a single category to build robust systems, refine their logistics, and understand their unit economics. Once that engine worked, they started to layer on new categories: first music and DVDs, then electronics, toys, and beyond, all riding on the same proven infrastructure.

Each expansion built on systems and economics they had already tested at scale. By the time Amazon pushed into a broader 'everything store,' they were not guessing how to fulfil orders or make the numbers work. They had a solid foundation — and they scaled on top of it.

The Scaling Readiness Scorecard

Score yourself honestly on each dimension.

The Scaling Readiness Scorecard

	Score (0-10 points each)
BUSINESS MODEL	
Unit economics are profitable (LTV is 3x+ CAC)	
Revenue model is repeatable and predictable	
Gross margins support growth investments	
OPERATIONS	
Core processes are documented and repeatable	
Quality stays consistent as volume increases	
Systems exist to track key metrics in real time	
TEAM	
Management layer can operate independently	
Hiring process is defined and working	
Culture is strong enough to absorb new people	
FINANCIALS	
12-18 months of cash runway	
Clear path to profitability at scale	
Financial reporting shows unit economics	
MARKET	
Market is large enough to support growth targets	
Competitive position is defensible	
Demand exceeds current capacity	
TOTAL SCORE	

Total Score	
40-50	You're ready to scale.
30-39	You're close. Fix the weak spots first.
20-29	You have work to do. Don't scale yet.
Below 20	Focus on building the foundation.

This scorecard works whether you're scaling a tech startup, a manufacturing operation, a service business, or a retail company. The specific metrics differ by industry, but the categories stay the same.

My Story: The Warning Signs We Missed

Looking back at my directory business, the warning signs were there; we weren't paying attention.

We had proven product-market fit. Our agents were making sales. We had the right pricing. The market existed. We weren't guessing about whether the business worked. It did.

But we needed more revenue. Our partner expected growth. Future investors would want to see scale. We had costs to cover. The path seemed obvious: hire more sales agents. We had four agents making sales. If four agents generate $X revenue, then 60 agents should generate 15X revenue. Simple math.

Except it wasn't.

We recruited aggressively. We went from four to 60 agents in a couple of months. We watched our total sales volume. We were looking for the overall number to climb. And it did climb, just not the way we expected.

What we failed to do was track individual agent performance. We didn't build systems to monitor each salesperson's conversion rates. We didn't have dashboards showing us who was performing and who wasn't. We assumed that if our original four agents could sell, the new 56 could too.

They couldn't. At least not at the same rate.

The average sales per agent dropped. Not a little. A lot. Our best agents kept performing. But the new agents weren't hitting anywhere near the same numbers. We had 15 times more people but nowhere near 15 times more productive output.

Our costs had scaled linearly - 60 salaries instead of four. But our revenue hadn't scaled the same way. We were burning through cash faster than we were bringing it in. Our cost-to-revenue ratio got worse, not better.

By the time we realized what was happening, we were weeks away from running out of money. We had to let 36 people go just to stabilize our cash flow.

The warning signs we missed weren't about product-market fit. We had that. They were about operational readiness.

We had built a system to track individual performance, but in our haste to scale, we took our eye off those numbers. We had created a training program that could replicate what our best agents did, but we never tested whether new hires were performing at the same level when selling to customers. We just hired bodies and hoped they'd be as good as the originals.

That's not scaling. That's wishful thinking.

If we had been watching the right metrics - individual conversion rates, not just total volume - we would have caught the problem in week two, not month three. We would have slowed down hiring. We would have focused on training and performance management. We would have built the systems we needed before we scaled.

Instead, we learned the expensive way: you can't scale what you can't measure, and you must monitor what you've decided is important enough to measure.

Moving Forward

Readiness isn't about perfection. No business is perfectly ready to scale. But there's a difference between "we have gaps to work on" and "we have no foundation at all."

The next chapters will walk you through building that foundation. The metrics you need to watch. The systems you need to build. The team you need to develop. The funding you need to secure.

But none of that matters if you start before you're ready. Take the time to assess honestly. Your future self will thank you.

2

The Metrics That Actually Matter

You can't manage what you don't measure. You've heard this a thousand times. It's still true.

But here's what nobody tells you: measuring the wrong things is worse than measuring nothing at all.

Wrong metrics give you false confidence. They tell you growth stories while you're actually bleeding out. They let you believe everything's working until the day the bank account hits zero.

I know this because I lived it. We measured total sales volume. That number went up. We felt good. What we didn't measure was individual agent performance. We didn't track our cost-to-revenue ratio. We didn't watch our cash burn per new hire.

Those were the numbers that mattered. We ignored them until it was almost too late.

The difference between companies that scale successfully and companies that implode mid-growth isn't luck. It's metrics. The successful ones watch the right numbers. They see problems early. They adjust before small issues become catastrophic.

The ones that fail? They optimize for metrics that feel good but don't mean anything.

The Vanity Metrics That Will Destroy You

Vanity metrics are numbers that look impressive but don't predict success. They make great headlines. They impress investors who don't know better. They feel like progress.

They're also useless.

Here are the most dangerous ones:

Total Users/Customers

A million users sounds amazing. A million users who don't pay, don't engage, and churn out in 30 days? That's a problem with a big number attached.

What matters isn't how many users you have. It's how many engaged, paying, retained customers you have. Quality beats quantity every single time.

Total Revenue

Revenue growth feels like success. You went from $100K per month to $500K per month. You're scaling, right?

Maybe. Or maybe your customer acquisition cost went from $50 to $400. Or maybe your churn rate has doubled. Or maybe you're losing money on every sale.

Revenue without profit is just expensive activity.

Page Views, Downloads, Signups

These are top-of-funnel metrics. They tell you people found you. They don't tell you if those people became customers. They don't tell you if those customers stayed. They don't tell you if you made money.

I've seen companies celebrate 100,000 downloads while their activation rate was 2% and their retention was near zero. Those downloads meant nothing.

Social Media Followers

Unless you're an influencer making money from attention, follower counts are meaningless. Ten thousand followers who don't buy anything contribute zero revenue.

The trap with vanity metrics is that they're easy to manipulate. You can buy users. You can discount your way to revenue. You can game algorithms for downloads. The numbers go up. You feel momentum.

Then you run out of money because none of it was sustainable.

Metrics Work the Same Way Across Every Industry

Before we get into which metrics matter, let's be clear about something: the principles in this chapter apply whether you're running a software company, a manufacturing operation, a professional services firm, a retail business, or a restaurant chain.

The specific numbers you track will look different. A SaaS business tracks monthly recurring revenue and user engagement. A manufacturing business tracks production output and inventory turns. A service business tracks billable hours and project margins. A restaurant tracks covers per night and food cost percentages.

But the underlying question is always the same: Do you make more money from each customer than it costs to acquire and serve them?

That's unit economics. And unit economics determine whether your business can scale profitably, no matter what industry you're in.

A furniture manufacturer needs to know the profit per piece and the cost to acquire each retail buyer. A consulting firm needs to know the per-project margin and the cost to win each client. A landscaping company needs to know the profit per job and the cost to acquire each homeowner contract.

The language changes. The calculations look different. But the discipline is identical.

If you can't answer the unit economics question for your business, you're not ready to scale. Full stop. It doesn't matter if you're selling software or sandwiches.

The Only Numbers That Tell the Truth

If vanity metrics are fool's gold, what's the real gold? Unit economics.

Unit economics answer one question: Do you make more money from each customer than it costs to acquire and serve them?

If yes, you have a business that can scale. If no, you have a problem that growth will make worse.

Here's what you need to know:

Customer Acquisition Cost (CAC)

This is every dollar you spend to acquire one new customer. Marketing costs, sales salaries, advertising, commissions, tools - everything divided by the number of new customers you got.

If you spent $10,000 on marketing and sales last month and acquired 50 customers, your CAC is $200.

Be honest about this number. Include all costs. Don't just count ad spend. Count salaries. Count software. Count everything that went into getting those customers.

Different industries have different CAC structures. A software company might spend heavily on online ads and inside sales. A manufacturing business might spend on trade shows and field sales. A service business might rely on referrals and networking. A restaurant might spend on local advertising and promotions. Calculate what it actually costs you.

Lifetime Value (LTV)

This is how much profit you make from a customer over their entire relationship with you.

For a subscription business, it's monthly revenue minus cost to serve, multiplied by the number of months they stay.

For a transaction business, it's average order value times purchase frequency times customer lifespan, minus costs.

For a service business, it's total contract value minus delivery costs.

For a manufacturer selling to retailers, it's the profit per order times the number of orders over the relationship.

LTV is harder to calculate early on because you don't have enough data on how long customers stay. Make your best estimate. Update it as you learn more. Just don't lie to yourself.

The LTV:CAC Ratio

This is the metric that matters most.

Divide your LTV by your CAC. If LTV is $600 and CAC is $200, your ratio is 3:1.

Here's what the ratios mean:

- **3:1 or better**: You're healthy. You can scale. You make three dollars for every dollar you spend acquiring customers.
- **2:1 to 3:1**: You're viable but tight. Scale carefully. Improve efficiency where you can.
- **1:1 or worse**: You're losing money on every customer. Do not scale. Fix the model first.

This ratio works across industries. A manufacturing business needs healthy margins per customer. A SaaS business needs strong retention. A service business needs repeat work. A retail operation needs customer frequency. The calculation looks different but the principle is the same: make more than you spend.

Payback Period

How long does it take to recover your CAC?

If your CAC is $200 and customers generate $50 in profit per month, your payback period is four months.

The shorter the better. Long payback periods mean you need tons of cash to fund growth. Short payback periods mean growth can fund itself.

Aim for under 12 months. Under six is even better.

Using These Metrics Together: The Unit Economics Health Matrix

Understanding LTV:CAC and payback period individually is useful. Understanding them together tells you what to do next.

Your dashboard tracks the numbers. This matrix turns those numbers into decisions.

To know where you stand, plot your business on two axes:

Vertical Axis: Your LTV:CAC Ratio (from 1:1 at the bottom to 5:1+ at the top)

Horizontal Axis: Your Payback Period (from 18+ months on the right to 3 months on the left)

The Unit Economics Health Matrix

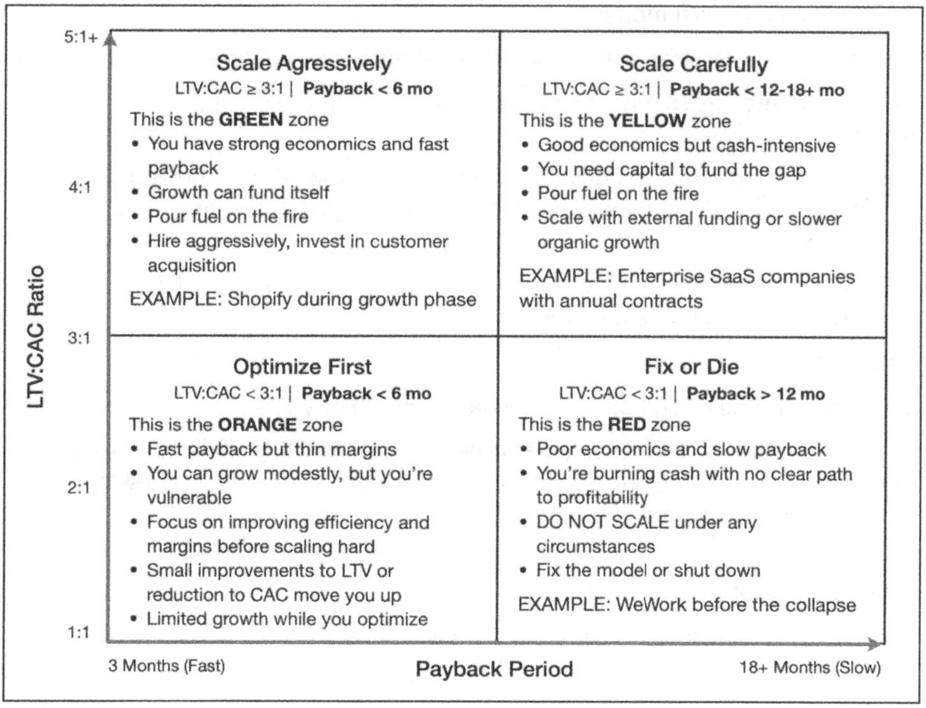

Where you land determines your strategy:

This matrix works across every industry. A manufacturing company plots their metrics the same way a software company does. A service business uses the same framework as a retail operation. The decision logic is universal.

We would have landed in the bottom right quadrant. Our CAC was climbing. Our LTV was shrinking because agents weren't converting. Our payback period was extending because new agents took longer to become productive. We were in "Fix or Die" territory, and we kept scaling anyway.

The matrix would have told us to stop. We hadn't paid attention to the data to plot ourselves on it.

Gross Margin

This is revenue minus the direct costs to deliver your product or service.

For software, gross margins can be 80-90% because delivery costs are tiny.

For manufacturing, gross margins might be 30-50% because materials and production costs real money.

For services, gross margins depend on how efficiently you deliver.

For restaurants, gross margins on food might be 60-70%, but total margins are much lower after labor.

You need strong gross margins to fund growth. If you're only keeping 20% after direct costs, you don't have much left to pay for sales, marketing, and operations.

Know your gross margin. Know what's normal for your industry. If yours is significantly lower, figure out why.

Churn Rate

What percentage of customers leave each month or year?

For subscription businesses, this is critical. A 5% monthly churn rate means you lose half your customers in a year. You're running on a treadmill. Growth just replaces lost customers.

For transaction businesses, churn looks different. It's about repeat purchase rates. Do customers come back?

For service businesses, it's about contract renewals and ongoing relationships.

For manufacturing, it's about reorder rates from your buyers.

Low churn is your secret weapon. If you keep customers, you can invest more to acquire them. If they leave quickly, you're stuck with low CAC limits.

Burn Rate and Runway

How much cash are you spending each month? How many months until you run out?

This is basic but critical. I watched our burn rate climb as we hired 60 agents. We didn't model it properly. We didn't calculate our runway. We found out the hard way.

Know your monthly burn. Know your runway. Track it weekly as you scale.

Essential Metrics by Stage

The metrics you watch change as you scale.

Early Stage (Pre-Scale)

Focus on proving the model:

- Product-market fit indicators (retention, engagement, NPS)
- Unit economics (LTV, CAC, payback period)
- Revenue per customer
- Gross margin

Don't worry about scale metrics yet. Worry about whether the model works at small scale.

Early Scaling (10-50 People)

Focus on consistency:

- LTV:CAC ratio across channels
- Churn rate by cohort
- Revenue per employee
- Cash burn and runway
- Individual contributor performance (like our agent conversion rates)

This is where we failed. We added people but didn't track individual performance. We watched total numbers, not per-person efficiency.

Growth Stage (50-200 People)

Focus on efficiency:

- Sales efficiency (revenue per sales rep, sales cycle length)
- Marketing efficiency (CAC by channel, conversion rates by stage)
- Operational efficiency (cost to serve, support tickets per customer)
- Team productivity metrics

Mature Scaling (200+ People)

Focus on optimization:

- Category leadership metrics (market share, brand awareness)
- Product expansion metrics (cross-sell, upsell rates)
- Profitability metrics (EBITDA, operating margin)
- Retention and expansion revenue

Industry affects which specific metrics matter most. A SaaS company obsesses over MRR and logo retention. A manufacturing business tracks inventory turns and production efficiency. A service business watches utilization rates and project margins. A retail operation monitors same-store sales and inventory turnover. But the core principle stays the same: measure what drives profitability.

Building Your Scaling Dashboard (And Actually Using It)

A dashboard nobody looks at is worthless.

The dashboard should fit on one screen or page. If you need 50 metrics to understand your business, you don't understand your business.

We didn't have this. We had spreadsheets. We had reports. But we didn't have one place where we could see the truth quickly. We watched what was easy to track (total sales) instead of what mattered (per-agent performance and cost ratios).

Build the dashboard before you scale. Update it religiously. Review it weekly with your team.

You need metrics that are:

- **Visible**: Everyone on the leadership team sees them
- **Current**: Updated weekly at minimum, daily for critical metrics
- **Actionable**: Tied to specific decisions you can make
- **Honest**: No gaming the numbers to look good

The Scaling Dashboard should use a single-page layout divided into five sections, each with visual health indicators. Each metric should show:

- Current number (large and bold)
- Small trend arrow (↑ improving, ↓ declining, → flat)
- Color indicator (Green/Yellow/Red based on health thresholds)

Here's what belongs on your scaling dashboard (a sample dashboard is on the next page:

The Scaling Metrics Dashboard

Top Banner - Company Snapshot (Horizontal across top)

- Revenue (current month + % growth vs. last month)
- LTV:CAC Ratio (with color coding: Green 3:1+, Yellow 2-3:1, Red below 2:1)
- Gross Margin %
- Cash Runway (in months, with warning indicator if under 6 months)
- Revenue per Employee

Below the banner, there are four equal-sized quadrants:

Quadrant 1: Revenue Metrics (Top Left)

- New Customers This Month
- Churn Rate (%)
- Monthly Recurring Revenue or Repeat Customer Rate (depending on business model)
- Average Contract/Order Value

Each with trend arrow and color coding

Quadrant 2: Unit Economics (Top Right)

- CAC by Channel (bar chart showing 3-4 main channels)
- LTV by Cohort (showing last 3-6 cohorts)
- Payback Period (in months)
- Customer Profitability (% of customers that are profitable)

Quadrant 3: Operations (Bottom Left)

- Key Performance Indicator #1 (varies by industry - could be sales conversion rate, production output, billable hours, etc.)
- Key Performance Indicator #2 (varies by industry)
- Customer Satisfaction Score or NPS
- Quality Metric (defect rate, support tickets, delivery time, etc.)

Quadrant 4: Cash (Bottom Right)

- Current Cash Balance
- Monthly Burn Rate
- Runway (redundant with top banner but critical enough to repeat)
- Days Sales Outstanding or Collections Rate

Visual Design Elements for the Dashboard:

- Use speedometer/gauge graphics for percentages and ratios (makes health immediately visible)
- Use small line graphs (sparklines) next to numbers to show 3-6 month trends
- Traffic light color coding throughout (Green = healthy, Yellow = caution, Red = danger)
- Keep it to one page - no scrolling required
- White space between sections for clarity

Adaptation for Different Industries

The quadrants stay the same, but the specific metrics in the "Operations Quadrant" change to reflect what's most critically important to the success of the business:

- **SaaS**: Active users, feature adoption rate, API uptime
- **Manufacturing**: Production efficiency, inventory turns, defect rate
- **Services**: Utilization rate, project margin, on-time delivery
- **Retail**: Same-store sales, inventory turnover, foot traffic conversion

Key Principle

The dashboard should answer three questions in under 60 seconds:

1. Are we healthy? (color coding tells you instantly)
2. Are we improving or declining? (trend arrows)
3. Where do we need to focus? (red/yellow indicators show problems)

This is your tracking tool. It shows you what's happening across all key dimensions of your business right now.

The Dashboard and the Unit Economics Health Matrix serve different purposes. The Dashboard tells you where you are. The Health Matrix tells you what to do about it.

- **Dashboard = "What's happening right now?"** (tracking tool, updated weekly)
- **Health Matrix = "What should we do about it?"** (decision tool, reviewed quarterly)

Use them together. Track your metrics weekly on the Dashboard. Plot your position quarterly on the Unit Economics Health Matrix. Adjust your strategy based on which quadrant you're in.

The Metrics You Inherit When You Acquire a Business

Buying a business instead of building one changes the metrics game.

You're not proving a model from scratch. You're evaluating a working model to decide if you can improve it.

Due Diligence Metrics

Before you buy, get crystal clear on:

- **Historical financials**: Three years minimum. Look for trends, not single years.
- **Customer concentration**: Do 80% of revenues come from three customers? That's risk.
- **Churn data**: Are customers leaving? Why?
- **Unit economics**: Does each customer generate profit?
- **Owner dependency**: Will revenue drop when the current owner leaves?

Many small business owners don't track these metrics well. That can be an opportunity. If you can bring data discipline to a business that's been running on gut feel, you can unlock value.

But you need to know what you're buying. Bad metrics hidden behind accounting tricks will destroy you.

Post-Acquisition Metrics

After you buy, watch:

- **Revenue retention**: Did customers stay when ownership changed?
- **Employee retention**: Did key people leave?
- **Integration costs**: What's it costing to merge systems and operations?
- **Improvement metrics**: Are the changes you're making working?

Acquisition can accelerate scaling if you buy well and integrate smartly. It can also be a disaster if you overpay for a business with broken metrics.

Case Study: Scaling on Solid Metrics - Shopify

Shopify grew from a small ecommerce software company into a global platform by obsessing over the right metrics, not just vanity numbers. Instead of focusing only on how many merchants signed up, they cared about how well those merchants performed and how long they stayed. In other words, retained, growing merchants were far more valuable than a constant stream of new signups that quickly churned.

They used revenue-based retention metrics — the same kinds of 'net revenue retention' figures they reported to public investors — to understand whether their existing merchants were expanding their sales over time, a practical proxy for merchant lifetime value. Their pricing model tied Shopify's revenue to merchant volume through subscriptions and a percentage of sales, so when merchants sold more, Shopify earned more too. That alignment made their unit economics fundamentally healthier.

Because they had data on merchant behavior, retention, and revenue, Shopify could justify investing aggressively in acquiring new merchants while still aiming for attractive lifetime value to acquisition cost ratios. Over many years, this focus on merchant success and economics allowed Shopify to scale rapidly while building a business model that could support long-term, sustainable growth.

Case Study: Growing Fast on Fake Numbers - WeWork

WeWork rocketed to a $47 billion valuation by showing impressive growth metrics: new locations opening, members joining, and revenue climbing. On the surface, it looked like hyper-growth done right.

Underneath, the unit economics were a mess. WeWork signed long-term, expensive leases and spent heavily to build out premium spaces, locking in huge fixed costs for years. Then it rented desks to startups and companies on short, flexible agreements they could cancel quickly. A mismatch like that is fragile: when demand wobbles, the revenue disappears, but the rent and fit-out costs don't.

WeWork poured money into expansion and discounts to fill buildings, yet most of the business still lost money. Margins were thin or negative, and there was no credible path to profitability. The company promoted highly adjusted numbers like 'community-adjusted EBITDA' that stripped out many real costs and made the losses look smaller than they were. Investors bought the story—until they read the IPO filings and focused on the real metrics.

Once the underlying economics were exposed, WeWork's valuation collapsed from $47 billion to a fraction of that within months, and it eventually ended up in bankruptcy.

That's not scaling; that's burning money at scale.

My Story: The Numbers We Watched (And the Ones We Ignored)

We watched the Total Sales volume. That was our north star metric. If total sales went up, we were winning.

It was the wrong metric.

Total sales did go up when we hired 56 more agents. Not as much as we hoped, but it went up. We celebrated that. We thought we were on track.

What we didn't watch:

Individual agent conversion rates: Our benchmark was simple: 1 sale per day or a conversion rate of 0.83%. Our original agents hit this consistently. We built our business model expecting 20-25% of new hires to underperform. That felt manageable.

As we scaled, the percentage of agents missing that benchmark exploded from 25% to 80%. Only 20% of our team was actually producing.

The problem? We prioritized hiring fast over hiring right. We tracked total sales instead of individual agent performance. By the time we looked at the individual numbers, we'd already built a team where most people couldn't do the job.

Cost per sale: With four agents on salary, our cost per sale was manageable. With 60 agents, our cost per sale quintupled. We didn't calculate this until we were in trouble.

Ramp time to productivity: We assumed new agents would perform immediately. They didn't. Some took months. Some never got there. We didn't measure how long it took to become productive.

Cash burn per new hire: Every new agent cost salary, training time, office space, and tools before they made their first sale. We didn't model this precisely enough. We just hired.

If we had tracked individual performance metrics, we would have known in week two that most new agents weren't performing. We would have stopped hiring. We would have focused on training and coaching. We would have built a system to replicate what our best agents did.

Instead, we kept hiring because total volume was the metric we watched. By the time we looked deeper, we were weeks from running out of cash.

The right metrics would have saved us. The wrong metric nearly killed us.

Moving Forward

Metrics aren't just numbers on a dashboard. They're your early warning system. They tell you what's working before it's obvious. They show you what's breaking before it's catastrophic.

Pick the right metrics for your business and stage. Build a dashboard you actually look at. Track individual performance, not just aggregate totals. Know your unit economics cold.

The next chapter will show you how to build the systems that improve those metrics. But systems without metrics are just activity. You need to know what to measure before you know what to build.

Get this right, and scaling becomes manageable. Get it wrong, and you'll grow yourself into bankruptcy.

3

Building Systems Before You Need Them

Systems feel like overkill when you're small. You've got four people. Everyone knows what to do. You figure things out as you go. Adding process feels like adding bureaucracy.

Then you scale and you discover something: having systems isn't enough. You also need the discipline to use them.

We had systems. We built a training program. We created sales scripts. We documented workflows. We did the work that most scaling companies skip.

Then we made a different mistake. We let people who didn't meet our standards onto the sales floor anyway.

After training, some new agents hit our performance benchmarks. They were ready. Others didn't. They struggled. They weren't converting calls to sales at acceptable rates.

We should have said no. We should have extended their training. Or we should have let them go. Instead, we hoped they'd improve once they were surrounded by successful team members. We told ourselves they just needed more time. We kept adding them to the team.

They never improved. The poor performers stayed poor performers. And because we took our eye off the key metric - sales per agent per day - we didn't see how badly it was affecting our overall numbers until it was too late.

We went from four high-performing agents to 60 agents with wildly varying performance. Our systems were fine. Our standards weren't.

This is the trap most founders miss. Everyone talks about building systems. Fewer people talk about having the discipline to enforce them. Systems tell you what good looks like. But you still have to make the hard decision to only accept good.

We didn't. It nearly destroyed us.

The hard truth about scaling is this: systems are necessary but not sufficient. You need systems to define standards. Then you need courage to maintain those standards when it's uncomfortable.

The Infrastructure Paradox

Here's the trap: build systems too early and you're drowning in process before you need it. Build them too late and you're drowning in chaos.

Companies that over-systematize early become rigid. They build elaborate processes for problems they don't have yet. They create org charts for imaginary teams. They implement enterprise software when they have five customers. The overhead kills them.

Companies that under-systematize scramble constantly. Everything is custom. Everything requires the founder. Nothing is repeatable. They can't scale because they can't let go.

The right time to build systems is just before you desperately need them.

That means different things by stage:

At 5-10 people: Document your core processes. How do you deliver your product or service? How do you onboard customers? How do you handle support? Write it down while it still fits on a few pages.

At 10-25 people: Build your first management layer. Create basic financial reporting. Implement light project management. Add structure without strangling flexibility.

At 25-50 people: Formalize hiring processes. Build real training programs. Create performance management systems. This is where we were when we scaled. We had these systems. We just didn't enforce the standards they defined.

At 50-100 people: Professionalize operations. You need real HR, real finance, real systems for everything. The founder can't be involved in every decision anymore.

Industry matters here. A software company can stay lean longer because digital systems scale cheaply. A manufacturing operation needs systems earlier because physical operations demand consistency. A service business falls somewhere in between.

But the pattern is universal: systems lag behind headcount. You build for where you're going, not where you are.

Which Systems Come First

You can't build everything at once. You'll exhaust yourself and your team. Prioritize based on what breaks first when you scale.

Operations Systems (First Priority)

These are the systems that deliver your product or service to customers. If these break, revenue stops.

For a software company: deployment processes, bug tracking, customer onboarding, feature development workflow.

For a manufacturing company: production schedules, quality control, inventory management, supply chain coordination.

For a service business: project scoping, delivery workflows, client communication, work product quality checks.

For a sales operation like ours, the processes were: lead qualification, sales scripts, objection handling, closing processes, and performance tracking.

Document these first. They're your revenue engine. If you scale without systematizing how you deliver, quality collapses and customers leave.

We had documented our sales process. Our scripts were good. Our workflow was clear. But we didn't enforce performance standards. We let underperformers stay on the team hoping they'd improve. The system was fine. Our discipline to use it wasn't.

Financial Systems (Second Priority)

You need to know if you're making or losing money. At scale, gut feel doesn't cut it.

Basic financial systems include:

- Monthly P&L by department or product line
- Cash flow tracking and forecasting
- Accounts receivable and payable management
- Budget vs. actual reporting
- Unit economics tracking (from Chapter 2)

You don't need enterprise-grade accounting software at 10 people. You do need it at 50. And you need someone who knows how to use it.

We tracked revenue and expenses. But we didn't track cost per agent rigorously enough. We didn't model cash flow with our hiring plans properly. We didn't have clear visibility into which agents were profitable and which weren't. When we started bleeding cash, we couldn't pinpoint the problem quickly enough.

People Systems (Third Priority)

As you add people, you need systems to hire, onboard, develop, and retain them.

Key people systems:

- Structured hiring process (job descriptions, interview frameworks, reference checks)
- Onboarding program (what every new hire learns in week one, week two, month one)
- Performance management (how you give feedback, set goals, track progress)
- Training and development (how people get better at their jobs)

This is where scaling companies live or die. If you can't bring new people up to speed quickly, you can't scale. If your best people leave because there's no growth path, you can't scale.

We had training. We had onboarding. What we didn't have was a clear performance gate. We didn't have a system that said "if you don't hit X conversion rate after Y training period, you don't move to the sales floor." We let people through who shouldn't have made it.

Build the system. Then build the discipline to enforce it.

Technology Systems (Fourth Priority)

Technology should enable your other systems, not replace them.

Common mistakes:

- Buying software before defining the process it should support
- Over-engineering solutions for simple problems
- Letting vendors tell you how your business should work

Smart technology choices:

- CRM for managing customer relationships and sales pipelines
- Project management tools for coordinating work
- Communication tools that don't create information overload
- Reporting tools that surface your key metrics

The technology should disappear. It should make work easier, not add an administrative burden.

Buy for your current scale, not your imagined future scale. You don't need Salesforce Enterprise at 10 people. You might not even need Salesforce. A spreadsheet might work fine.

Documenting Your Magic Before It Breaks

Every successful small business has magic. The thing that makes you different. The approach that wins customers. The quality that sets you apart.

That magic usually lives in the founder's head. Or in your best employees' habits. It's not written down. It's not teachable. It's just how things are done.

This works until you scale. Then the magic dilutes. New people don't have it. They do things their own way. The results aren't the same. Quality slips.

Before you scale, document the magic:

What do your best performers do differently? Shadow them. Record them. Interview them. Figure out what makes them successful. Turn it into a process others can learn.

We did this. We documented what our top four agents did. We refined that as our agents improved their selling rates. We built training around it. But documentation isn't enough. You also need selection criteria. Not everyone can execute at the level you need, even with good training. You need to identify who can and who can't, then make the hard call.

What makes customers choose you? Is it speed? Quality? Expertise? Personal attention? Whatever it is, build systems that preserve it at scale.

A consulting firm's magic might be their discovery process. A manufacturer's magic might be their quality control. A software company's magic might be their customer support. A restaurant's magic might be their consistency.

Don't scale until you've captured and systematized what makes you special. And don't scale with people who can't deliver that magic, even with your systems.

Process Without Bureaucracy

Systems get a bad reputation because most people build them wrong. They create processes that slow everything down. Forms that nobody reads. Approvals that add no value.

Good systems make work easier. Bad systems make work harder.

Here's how to build systems that scale without creating bureaucracy:

Keep them simple: If you need a manual to understand the process, it's too complex. Simple systems get followed. Complex systems get ignored.

Make them relevant: Every system should solve an actual problem you're experiencing. Don't build systems for theoretical future problems.

Build in feedback loops: Systems should improve over time. Ask the people using them what's working and what's not. Update regularly.

Default to trust: Don't build systems that assume people will screw up. Build systems that assume people will succeed and help them do that.

Automate the tedious parts: If a system involves copying data between spreadsheets or filling out the same form repeatedly, automate it.

The goal isn't control. The goal is consistency and quality at scale.

What You Get (And Don't Get) When You Buy a Business

Acquisition entrepreneurship changes the systems equation.

When you buy a business, you inherit their systems. Sometimes that's the value - they've built infrastructure you don't have to create. Sometimes it's the problem - their systems are outdated or broken.

What you typically get:

- Established processes for operations and delivery
- Existing financial and accounting systems (though often dated)
- Some level of documentation (though rarely comprehensive)
- Technology systems (though often legacy)

What you typically don't get:

- Modern, efficient processes
- Well-documented procedures
- Integrated technology
- Systems that were built for scale

Many small businesses run on founder knowledge and muscle memory. The systems exist but they're not written down. When you buy the business, you're betting you can formalize and improve what's there.

The integration challenge is real. You have your systems, and they have theirs. Merging them smoothly without disrupting operations requires careful planning.

Ask these questions before you buy:

- Are key processes documented?
- Will the systems work without the current owner?
- What's the technology debt?
- How much would it cost to modernize?

Sometimes buying a business with weak systems presents an opportunity. You can introduce operational discipline and unlock value. However, it can also be a trap, as the business only works because the owner holds it together.

Know which one you're buying.

The Systems Priority Matrix

Not all systems are equally urgent. Some require immediate attention, while others can afford to wait.

To know which systems you need to work on urgently, plot them on a Systems Priority Matrix for your business.

Plot two axes:

Vertical Axis: Impact on Revenue (Low to High), and
Horizontal Axis: Current Pain Level (Low to High).

Then add your business's systems to one of four quadrants: **Build Now**, **Build Next**, **Optimize Later**, and **Ignore**.

A sample Matrix is below.

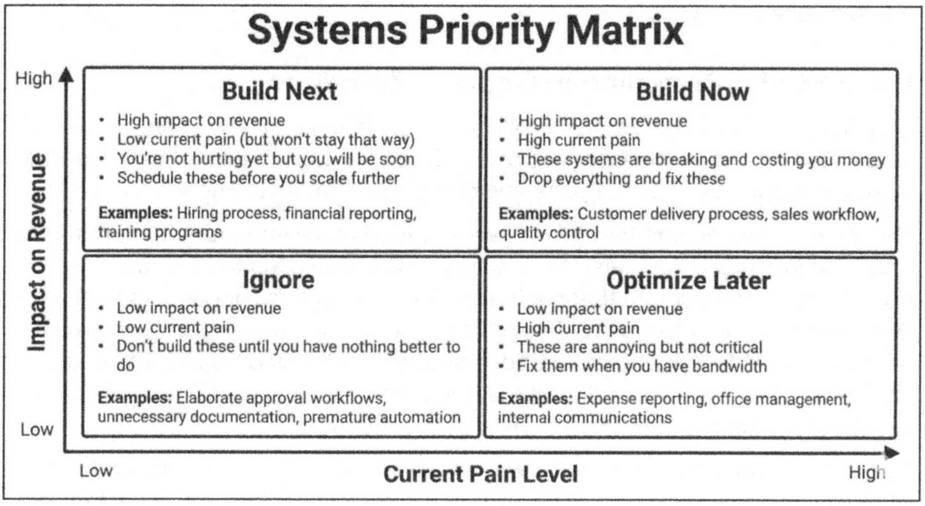

We had the training system. What we needed but didn't have was a performance-tracking and enforcement system. That was top right - high impact, high pain. We were losing money every day because low performers stayed on the team.

We needed a dashboard showing each agent's daily performance. We needed clear benchmarks. We needed the discipline to act when people didn't meet them.

Case Study: Building Just in Time - Stripe

Stripe scaled from a small startup to a payments giant by building systems deliberately.

In the early days, Patrick and John Collison did almost everything themselves. They wrote the core product, worked directly with early customers to help them integrate Stripe, and personally handled feedback and support. That closeness gave them a deep understanding of what was working and where the product broke.

As Stripe grew, they started to systematize methodically. They built internal tools to run operations, created repeatable processes for onboarding new customers, and invested heavily in documentation and engineering discipline. Instead of replacing speed with bureaucracy, they designed systems that made it easier for teams to move quickly while keeping quality and reliability high.

The result was a company that could process billions of dollars in payments while maintaining a strong reputation for reliability, developer experience, and service. Stripe didn't just grow fast — it grew fast on the back of systems that made scaling sustainable.

Case Study: Drowning in Chaos - Zenefits

Zenefits grew from zero to a $4.5 billion valuation in just a few years. By 2015, it was hailed as one of the fastest-growing SaaS startups in Silicon Valley, adding customers and revenue at a pace few companies had ever seen.

But the systems behind that growth were dangerously immature. Zenefits operated in a highly regulated insurance market and failed to build the compliance infrastructure that such a business demands. Regulators later found that unlicensed employees had been selling insurance and that staff were using an internal tool to bypass mandatory pre-licensing training hours. The culture and operations were chaotic, and the company was effectively trying to run a regulated brokerage with startup-style 'move fast' habits.

Regulators eventually hit Zenefits with multiple enforcement actions and millions of dollars in fines across several states. Customers lost trust, revenue targets were missed, and the board forced out the founder and CEO, Parker Conrad. The company's valuation and reputation collapsed, and Zenefits spent years restructuring under new leadership before ultimately being acquired.

They had the customers. They had the top-line growth. What they didn't have were the systems and controls required to deliver that growth safely at scale.

My Story: The System We Had and the Discipline We Didn't

Looking back, we did a lot right. We built a training program before we scaled. We documented our sales process. We created scripts and workflows. We had the infrastructure.

Our failure wasn't the system. It was our refusal to enforce the standards the system defined.

After training, we could see who was performing and who wasn't. The data was clear. As we hired more people and scaled, some agents hit acceptable conversion rates. Others didn't come close.

We should have enforced: "If you don't hit X conversions per Y calls after two weeks on the floor, you're not ready. We'll give you more training, or we'll part ways."

Instead, we said: "Keep trying. You'll get better. Watch the successful agents. Learn from them."

They didn't get better. They stayed on the team for months, generating minimal revenue while consuming salary, space, and management time.

We took our eye off daily sales per agent. We watched the total sales volume instead. As long as the total number was growing, we felt okay. We didn't see that we were growing the team faster than we were growing revenue.

By the time we accepted that we had too many underperformers, we were burning cash at an unsustainable rate. We had to let 36 people go just to survive.

The lesson isn't "build better systems." We had decent systems. The lesson is "have the courage to enforce your standards." Systems are worthless if you don't use them to make hard decisions.

If an agent wasn't hitting benchmarks after training, we should have faced it immediately. Not hoped. Not waited. Not made excuses. Faced it and acted.

That's what scales. Not just the systems, but the discipline to use them.

Moving Forward

Systems aren't sexy. They're not the vision. They're not the product. They're not what gets you excited about your business.

But systems are what let you scale without breaking.

Start documenting now. Build the critical systems before you desperately need them. Keep them simple. Make them useful. Update them as you grow.

And when your systems tell you something uncomfortable - when the data shows someone isn't performing, when the metrics show a process isn't working - have the discipline to act on it.

The next chapter will show you how to scale your team within these systems. But systems only work if you're willing to enforce the standards they define.

Build the systems. Then use them.

4

Scaling Your Team Without Losing Your Culture

Here's the fear that keeps founders awake at night: You'll scale to 100 people and won't recognize your own company.

The scrappy team that moved fast and cared passionately becomes a corporate machine full of people who just collect paychecks. The values that made you unique are replaced by bureaucracy and politics. The culture that attracted your best people evaporates.

It happens frequently. Small companies with fantastic cultures scale up and lose their spirit. The growth goes ahead, but the company dies from the inside.

The question isn't whether culture will change when you scale. It will. The real question is whether you can hold on to what matters while letting go of what doesn't.

Most founders get this wrong. They try to freeze culture in place, fighting every change. They insist that the company at 50 people should feel exactly like it did at 5. But this approach doesn't work. You can't scale nostalgia.

Or they ignore culture completely. They focus on growth and assume culture will take care of itself. It won't. Culture dilutes by default. If you're not actively preserving it, you're passively destroying it.

The companies that scale successfully do something different. They pinpoint the core elements that define their culture. Then they guard those elements fiercely while allowing everything else to evolve.

What Culture Actually Means (And What It Doesn't)

Culture isn't ping pong tables and free snacks. Culture isn't casual dress codes or Friday happy hours. Those are perks. Perks are easy. Culture is hard.

Culture is how decisions get made when the founder isn't in the room. Culture is what behavior gets rewarded and what gets punished. Culture is the unwritten rules that everyone follows because that's just how things work here.

In practical terms, culture shows up in:

How you treat customers: Do you go the extra mile or do the minimum? Do you own mistakes or deflect blame? Do you prioritize long-term relationships or short-term revenue?

How you treat each other: Do you give direct feedback or talk behind backs? Do you collaborate or compete internally? Do you help teammates succeed or protect your territory?

How you make decisions: Do you move fast and accept mistakes or move slow to avoid risk? Do you trust people or micromanage them? Do you debate openly or defer to hierarchy?

What you celebrate: Do you reward individual heroics or team success? Do you celebrate hitting numbers or delighting customers? Do you promote based on results or relationships?

What you won't tolerate: Where are your lines? What behavior gets someone fired immediately? What compromises will you never make?

These patterns define your culture. And when you scale, these patterns are what you need to protect.

Our culture at four agents was scrappy and performance-driven. If you sold, you succeeded. If you didn't, you didn't last long. Simple. Clear. Effective.

When we scaled to 60, we lost that clarity. We started tolerating poor performance. We hoped underperformers would improve instead of holding the line. We diluted the performance-driven culture that made us successful.

New hires saw that poor performance was tolerated. They adjusted their effort accordingly. The high performers got frustrated. The culture shifted from "perform or leave" to "hang around and collect a paycheck."

That killed us.

The Transition Points: What Changes as You Grow

Culture evolves differently at different scales. Understanding these transitions helps you know what to preserve and what to let go.

5 to 15 People: The Founding Team

At this size, culture is effortless. Everyone knows everyone. Communication is direct. You make decisions over lunch. The founder sets the tone by example.

What works:

- Informal communication
- Founder-led everything
- Hiring people you know
- Minimal process

What you need to start documenting:

- Your core values (even if they're obvious)
- The behaviors you reward
- The standards you won't compromise

15 to 50 People: The First Layers

This is where culture starts to fracture if you're not careful. You can't know everyone personally anymore. You need managers. Communication gets harder.

What changes:

- The founder can't be in every conversation
- Departments start to form
- Some people have never met each other
- Stories about "the early days" start circulating

What you must do:

- Hire managers who embody your culture
- Codify your values explicitly
- Create rituals that reinforce what matters (all-hands meetings, recognition programs, team events)
- Fire fast when someone violates core values

This is where we failed. We hired 56 agents in a few months. We didn't reinforce our performance culture. We didn't create rituals around celebrating high performers. We just added bodies and assumed the culture would scale with them.

It didn't.

50 to 150 People: The Professional Transition

At this scale, informal doesn't work anymore. You need real HR. Real processes. Real structure. Many founders hate this phase. It feels corporate. It feels like you're losing what made you special.

You're not. You're just growing up.

What changes:

- Formal policies become necessary (handbook, benefits, performance reviews)
- Culture gets codified in writing
- Onboarding becomes critical (new hires learn culture from a program, not osmosis)
- The founder becomes more symbol than daily presence

What you must do:

- Write down your values and what they mean in practice
- Build an onboarding program that teaches culture explicitly
- Promote and fire based on cultural fit as much as performance
- Tell the founding stories repeatedly (these become your mythology)

150+ People: The Scaling Organization

Culture becomes self-sustaining or it dies. You can't personally influence everyone anymore. Middle management carries the culture. New hires learn from people who learned from people who learned from you.

What works:

- Strong middle management who model values
- Clear, simple value statements everyone can remember
- Consistent consequences for cultural violations
- Regular reinforcement through communication

What kills culture:

- Tolerating "brilliant jerks" who hit numbers but violate values
- Promoting people who deliver results through toxic methods
- Saying values matter but rewarding behavior that contradicts them
- Letting different departments develop conflicting cultures

The size at which you hit these transitions varies by industry. A tech company might stay informal longer. A manufacturing operation might need structure earlier. A service business falls somewhere in between. But the pattern holds: what worked at one scale won't work at the next.

Hiring for Scale: Different Skills, Same Values

At four people, you hire for versatility. You need people who can do everything. At 60 people, you hire specialists. You need people who are excellent at specific things.

The skills change. The values can't.

This is the trap: you start hiring for credentials instead of fit. You hire the experienced VP because they've "done it before" at a big company. You hire the superstar individual contributor who hits numbers but treats people poorly. You hire quickly because you need bodies.

You compromise on values. The new hires don't match your culture. Over time, they become your culture.

How to Hire for Values:

Define what you actually value: Not what sounds good in marketing copy. What do you really care about? Write specific behaviors that demonstrate each value.

If "customer obsession" is a value, what does that look like? Does it mean responding to support tickets in under an hour? Does it mean calling customers after they churn to learn why? Get specific.

Interview for values, not just skills: Ask behavioral questions that reveal values. "Tell me about a time you had to choose between hitting a deadline and doing quality work. What did you choose and why?"

Listen for alignment with how you'd want people to decide.

Check references on culture, not just performance: Don't just ask "Did they hit their numbers?" Ask "What was it like to work with them? How did they treat people when things went wrong? What did they do when they disagreed with a decision?"

Make values a veto: If someone is skilled but doesn't match your values, don't hire them. Brilliant jerks destroy culture faster than mediocre people who fit.

We failed this completely. We were so desperate to fill seats that we hired anyone who could speak on the phone. We didn't screen for our performance culture. We didn't ask "Are you self-motivated? Can you handle rejection? Can you hit a sales target number? Do you need a lot of management or can you drive yourself?"

We hired people who needed constant supervision in a culture that rewarded independence. They never fit. They never would have fit. We should have seen it in the interview.

When Founders Need to Step Back (And When They Absolutely Shouldn't)

One of the hardest transitions in scaling is learning what you can delegate and what you must keep.

What founders must delegate:

Daily operations. You can't be in every meeting, approve every decision, review every output. You need managers who can run things without you.

Execution. You set the strategy. Other people execute it. If you're still doing the work instead of leading the people doing the work, you're the bottleneck.

Specialization. As you scale, you need experts who know more than you about their domains. Let them be experts. Don't micromanage engineering, finance, or HR when you hired people who know those functions better than you.

What founders must never delegate:

Culture. You can hire a head of HR to build programs. But the founder sets the tone. The founder decides what behavior is acceptable. The founder fires the high performer who violates values. This cannot be delegated.

Vision. Other people can build the roadmap. But the founder owns where the company is going and why it matters. If you delegate this, you're not leading anymore.

Key hires. Especially in early scaling. Every hire at this stage shapes culture. The founder must be involved in hiring until you have leaders who understand what you're building.

We delegated too much. We hired managers to run their sales team without ensuring they understood and would enforce our performance culture. They tolerated mediocrity because we tolerated mediocrity. If we had stayed directly involved in setting performance standards and making decisions about who stayed on the team, we wouldn't have diluted our culture.

The Delegation Dilemma

Founders struggle with delegation for two reasons:

You're afraid people will screw it up: They might. They will, actually. Everyone screws up. The question is whether you let them learn from mistakes or whether you take everything back.

If you take everything back, you become the bottleneck. Nothing scales. People stop trying. You burn out.

You define yourself by doing the work: You started this company. You built it with your hands. Letting go feels like losing your identity.

Get over it. Your value at scale isn't doing the work. It's building the organization that does the work. If you can't make that shift, you can't scale.

Here's how to delegate without losing control:

Define the outcome, not the process: Tell people what success looks like. Let them figure out how to get there. Judge them on results, not methods.

Create decision-making frameworks: Give people principles to guide decisions. "We always choose customer delight over short-term profit" is a framework. People can make decisions using that framework without asking you.

Build in reviews, not approvals: Instead of approving everything before it happens, review what happened and give feedback. This lets people move fast while ensuring quality.

Accept that 80% done by someone else is better than 100% done by you: If you insist on perfection, you'll never delegate. Let people do good work their way instead of perfect work your way.

Culture Integration When You Acquire

Buying a business brings a different challenge: merging two cultures without destroying both.

Every business has a culture. When you acquire a company, you're buying their people and their ways of working. Imposing your culture by force destroys value. Ignoring cultural integration creates chaos.

The Integration Spectrum:

Full Integration: You merge the acquired company completely into yours. Same systems, same values, same culture. Use this when you're buying a small company to fold into yours or when their culture is toxic and must change.

Partial Integration: You align on core values and systems but let them maintain some independence. Use this when you're buying capability or market presence and want to preserve what makes them effective.

Autonomous: You let them run independently with light coordination. Use this when you're buying a successful company that works because of their culture, and changing it would kill the value.

Questions to ask before you integrate:

- What about their culture creates value? (Don't destroy this)
- What about their culture conflicts with yours? (Must change)
- What about their culture is just different? (Can coexist)

Integration approach:

Start with values. Get leadership from both companies in a room. Discuss what each company values. Find the overlap. Be explicit about conflicts. Decide which values win.

Communicate clearly. Tell people from both companies what's changing and what's not. Don't pretend everything will stay the same. Don't pretend everything must change.

Keep the best of both. If they have a customer service approach that works better than yours, adopt it. If you have a performance management system that's stronger than theirs, implement it. Integration isn't conquest.

Move deliberately. You can't merge cultures overnight. Give people time to adjust. Expect friction. Work through it.

The Culture Preservation Framework

When scaling or integrating cultures, use this framework:

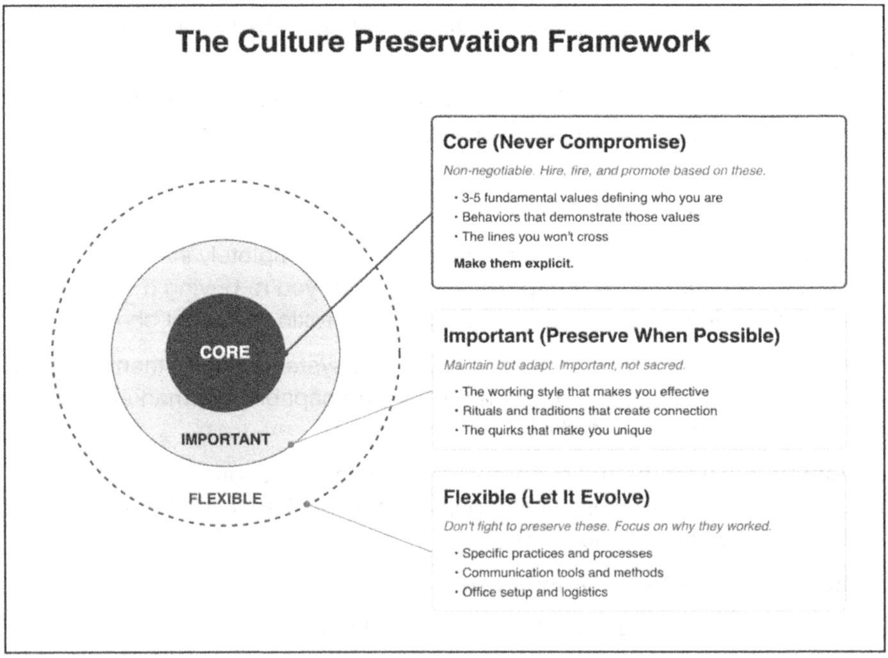

Core Values (Never Compromise)

Write these down. Make them explicit. Hire, fire, and promote based on them. These are non-negotiable.

Important Values (Preserve When Possible)

Try to maintain these. But be willing to adapt them as you grow. They're important but not sacred.

Flexible Values (Let It Evolve)

Let these change. Don't fight to preserve how things used to work. Focus on why they worked.

The mistake founders make is putting too much in "Core." They try to preserve everything. Culture becomes rigid. Scaling becomes impossible.

Or they put too much in "Flexible." They change everything. New hires don't know what the company stands for. Culture disappears.

Get the Core right. Protect it fiercely. Let everything else adapt.

Case Study: Scaling While Keeping Soul - Patagonia

Patagonia grew from a small climbing equipment company into a multi-billion-dollar outdoor apparel and gear business, with thousands of employees across many locations. Through that growth, it became known for something rare: it scaled without losing its soul.

Environmental responsibility and product quality sit at the core of Patagonia's identity. Those values are non-negotiable. The company has repeatedly made business decisions based on environmental impact, even when that meant higher costs or slower growth, and it refuses to compromise on making durable, high-quality products.

Patagonia hires people who share its mission and builds policies and practices —like support for activism, flexible work, and long-term environmental commitments—that reinforce those values every day. When actions or decisions conflict with those values, leadership has shown a willingness to walk away from revenue or change course rather than dilute the mission.

Everything else has been flexible. Processes have evolved, the organization structure has changed, and the markets Patagonia serves have expanded, but its core values have remained the reference point.

The result is more than 50 years of growth and strong performance, with one of the most distinctive and enduring cultures in modern business.

Case Study: Losing Culture Through Scaling - Yahoo

Yahoo grew from a scrappy startup into one of the internet's first great giants, hiring thousands of people and operating at massive scale. Over time, though, the company became bureaucratic, political, and slow.

One of the deeper problems was cultural. Former employees and observers argue that Yahoo never clearly defined and reinforced a strong core identity. Instead of hiring and rewarding people primarily for cultural fit and shared mission, it often rewarded credentials and status. Inside the company, teams competed for resources, built their own fiefdoms, and pushed conflicting priorities, making it hard to align around big new product bets.

As the politics and bureaucracy grew, morale fell and talented people left for more focused, faster-moving companies. Innovation slowed, and Yahoo gradually shifted from defining the consumer internet to struggling to stay relevant while rivals like Google and Facebook surged ahead.

Growth without cultural discipline doesn't just create friction. Over time, it can hollow out a company from the inside.

Case Study: Integration Done Right - Disney and Pixar

When Disney acquired Pixar in 2006, they faced a major integration challenge. Disney was a traditional media giant with a classic studio culture. Pixar was a creative-driven animation studio with a distinctive, highly collaborative culture. Disney's leadership recognized that Pixar's creative culture was a huge part of what made the studio valuable. Instead of forcing Pixar to conform to Disney, they kept Pixar as an autonomous studio and deliberately protected its way of working.

At the same time, Disney learned from Pixar's creative process and applied key ideas—like giving filmmakers more control and fostering candid, iterative feedback—to its own animation group. The two companies aligned around shared values of storytelling excellence while integrating distribution, business, and franchise systems. Pixar kept its creative identity; Disney plugged Pixar's strengths into its global machine.

The result: both sides got stronger. Pixar continued to produce hit films with its own culture intact, and Disney's animation arm was revitalized by adopting elements of Pixar's approach.

My Story: How We Diluted Performance Culture

Our culture with four agents was clear: performance matters most. If you sell, you're valued. If you don't sell, you don't last. Everyone knew this. It created focus and urgency.

When we scaled to 60, we lost that clarity.

We started accepting agents who didn't perform at our standard. We told ourselves they'd improve. We made excuses. We hoped surrounding them with high performers would elevate them.

The high performers saw us tolerating mediocrity. They wondered why they should work so hard when others could coast. Some stayed and lowered their effort. Some left for companies with higher standards.

The low performers saw they could survive without hitting benchmarks. They stuck around. They collected paychecks. They filled desks.

Our performance culture died not because we stopped caring about performance, but because we stopped enforcing standards. We said performance mattered. We rewarded people who didn't perform. Our actions contradicted our values.

New hires learned the real culture fast: the stated values don't matter. What matters is just showing up.

By the time we realized what happened, we had diluted our culture so much we had to fire 36 people to restore it.

The lesson: culture is what you do, not what you say. If you say performance matters but keep poor performers, your culture is "mediocrity is fine." If you say customer service matters but don't fix problems, your culture is "customers don't really matter."

Your team watches what you do. They ignore what you say. Make sure your actions match your stated values.

Moving Forward

Scaling your team without losing culture is possible. But it requires clarity and discipline.

Get clear on your core values. Write them down. Make them behavioral. Hire for them. Fire when people violate them.

Let everything else evolve. Your processes will change. Your structure will change. Your practices will change. That's fine. That's growth.

But protect your core. Make decisions that align with it, even when they're expensive or painful. The culture you preserve through scaling becomes your competitive advantage.

The next chapter will address how to fund that scaling without losing control. But funding without culture just gives you capital to build something you don't recognize.

Get the culture right first. The rest follows.

5

The Money Question

Funding Your Scale-Up

Money doesn't solve scaling problems. But lack of money will kill you while you're trying to solve them.

This is the paradox of funding: you need capital to scale, but capital comes with strings attached. Take too little and you run out of resources before reaching your goals. Take too much and you risk diluting yourself or losing control. Take it from the wrong people and you spend more time managing investors than building your business.

Most founders leave their funding too late. They're six months away from running out of cash before they start looking for money. By then, they're desperate. Desperation leads to bad deals.

Or they think about funding too early. They raise money before they know what they'll use it for. They scale prematurely because they have capital burning a hole in their pocket. The money enables bad decisions.

The right time to think about funding is before you start scaling. Know your options. Know what each path costs. Know what you're optimizing for. Then make a choice that aligns with what you're building.

Bootstrap vs. Raise: The Real Trade-Offs

The startup world glorifies venture capital. Raise money. Scale fast. Dominate the market. Anything else is playing small.

This is nonsense.

The vast majority of successful businesses never raise institutional capital. They bootstrap. They grow on revenue. They maintain control. They build slower but they build on their own terms.

Neither path is better. They're different. They optimize for different outcomes.

Bootstrapping means:

You own your company. Nobody tells you what to do. You make decisions based on what's right for the business, not what investors want.

You scale at the pace revenue allows. This can be slower. But it forces discipline. You can't outrun your ability to deliver.

You focus on profitability early. You have to. There's no safety net. This makes you efficient.

You answer to customers, not investors. If customers are happy and paying, you're winning. That's the only metric that matters.

The downside:

Growth is limited by cash flow. If you're in a market where speed matters, you might lose to funded competitors.

You can't make big bets. You can't invest heavily in R&D or marketing if it doesn't pay back quickly. You have to be conservative.

Personal financial pressure. If you're not taking a salary to preserve runway, that's stress.

Raising capital means:

You can move faster. You can hire ahead of revenue. You can invest in systems and technology. You can enter markets before they're profitable.

You can survive early losses. Funded companies can lose money for years while they build. Bootstrapped companies can't.

You get strategic help. Good investors bring networks, advice, and credibility.

The downside:

You give up ownership. Investors own a piece of your company. The more you raise, the less you own. Eventually, you might own a minority of what you built.

You answer to investors. They have expectations. Growth targets. Board seats. Reporting requirements. You're not fully in control anymore.

You're on a timeline. Investors expect returns. That means growth, exit, or IPO. You can't build a lifestyle business with venture capital. You're building to sell or scale massively.

The pressure changes. It's not "build a great business." It's "hit the numbers we projected." That changes how you make decisions.

Neither is right for everyone.

Bootstrap if:

- You value control over speed
- Your market doesn't require winner-takes-all speed

- You can reach profitability without massive upfront investment
- You want to build a sustainable business, not chase an exit

Raise if:

- Your market has network effects where first mover wins
- You need significant capital to reach minimum viable scale
- You're comfortable with outside accountability and reduced ownership
- You're building for a large exit

We bootstrapped initially. We had enough revenue from our first four agents to be sustainable. When we decided to scale to 60, we should have raised capital or scaled much more slowly. Instead, we tried to bootstrap aggressive growth. We didn't have the cash buffer to support that pace.

When to Raise Capital (And How Much)

If you decide to raise, timing and amount matter enormously.

When to raise:

After you've proven the model: Investors want to see traction. Revenue. Customers. Proof that your business works. The earlier you raise, the more expensive it is in terms of dilution.

Before you desperately need it: Start fundraising when you have 12-18 months of runway left, not when you have 3 months. Desperation shows. You'll take bad deals.

When you have a specific use for the capital: "We're raising to fund growth" is vague. "We're raising to hire 10 sales reps over 6 months based on a proven CAC and LTV model" is specific. Investors fund specific plans.

How much to raise:

Most founders raise too little or too much.

Too little means you run out of money before hitting your next milestone. You're back fundraising in 6 months when you should be building. You spend all your time raising instead of operating.

Too much means you have capital you don't need. It enables sloppy spending. You hire too fast. You lose discipline. And you've diluted yourself unnecessarily.

The right amount:

Calculate what you need to hit your next major milestone, plus a 6-month buffer. That's your target.

If you need to hire 20 people to reach $5M in revenue, model the cost. Salaries, space, equipment, marketing, buffer for mistakes. Add 20% contingency. That's your raise amount.

From whom to raise:

Friends and family: Easiest to raise from. Hardest to manage if things go wrong. Only take money from people who can afford to lose it.

Angel investors: Individuals investing their own money. Often entrepreneurs who've exited. Can provide advice and networks. Smaller check sizes mean more people to manage.

Venture capital: Professional investors managing funds. Larger checks. Higher expectations. More structured. Only relevant if you're building for massive scale.

Strategic investors: Companies in your industry. Can provide partnership opportunities. Can also complicate things if they're competitors or have conflicting interests.

Match your investor type to your stage and goals.

The Acquisition Financing Path: A Different Route to Scale

Raising capital to build is one path. Using financing to buy existing businesses is another.

This is acquisition entrepreneurship's secret weapon: you can finance acquisitions with minimal personal capital.

SBA Loans:

In the USA, the U.S. Small Business Administration (SBA) guarantees loans for business acquisitions. You can buy a profitable business with as little as 10% down. The business's cash flow services the debt.

Example: You want to buy a $2M business generating $400K in annual profit. You put down $200K. An SBA loan covers $1.8M. The business generates enough cash flow to pay the loan and still provide you with income.

You just acquired a profitable business with $200K instead of $2M. That's leverage.

Seller Financing:

Many business sellers will finance part of the purchase price. They want to sell. They want to see the business succeed under new ownership. They're willing to let you pay over time.

Example: You buy a $1M business. The seller finances $300K over 5 years. You need $700K upfront instead of $1M. And the seller stays motivated to help you succeed because they haven't been paid in full yet.

Earnouts:

You structure the deal so that part of the purchase price depends on future performance. If the business hits certain targets post-acquisition, the seller gets paid more.

This aligns incentives. The seller wants you to succeed. You reduce upfront cost. You pay more only if the business performs.

Why this matters for scaling:

Instead of building slowly from zero, you buy a $2M revenue business today. You bring operational improvements. You scale it to $5M. You've just scaled faster and with less risk than starting from scratch.

The capital requirements are different. You need to qualify for loans. You need to understand deal structures. But you don't need venture capital. You don't give up equity. You maintain control.

Managing Cash Flow During Rapid Growth

Here's what nobody tells you: scaling burns cash.

You hire people before they're productive. You invest in systems before they pay off. You build inventory before you sell it. You extend payment terms to win customers while paying your own bills on time.

This is the cash flow crunch. Revenue is growing. You're winning. But your bank account is shrinking.

Companies that don't model this carefully run out of cash in their best growth periods. They're succeeding themselves to death.

The mechanics of the cash crunch:

You hire someone today. They cost salary immediately. They won't be productive for 2-3 months. During their ramp, they're pure cost.

You win a customer. Great. But they pay in 30-60 days. You pay your team weekly. You're funding 2 months of operations before you collect revenue.

You buy inventory or materials. You pay now. You sell later. The gap consumes cash.

You invest in marketing. You pay today. Customers show up next quarter. Your CAC comes before your LTV.

This is normal. This is scaling. But if you don't model it, you'll wake up one day with no cash and wonder how you got there.

How to manage it:

Model your cash flow with your growth plans: Before you hire 56 people like we did, model the cash impact. When do they start? When do they become productive? What does revenue look like in months 1, 2, 3, and 4?

We didn't do this. We modeled revenue optimistically. We didn't model the ramp time. We didn't model the cost of underperformers. By the time we realized we were burning cash faster than we were generating it, we had 8 weeks left.

Build a cash buffer: You need 6-12 months of runway when you start scaling. Not 2-3 months. Scaling takes longer than you think. Problems cost more than you expect. Without a buffer, you're one bad month from disaster.

Watch cash daily: Revenue is a lagging indicator. By the time revenue drops, you're in trouble. Cash is real-time. Track it daily. Know your burn rate. Know your runway. Update weekly.

Extend your payables, shorten your receivables: Negotiate payment terms with vendors. Push them out. Incentivize customers to pay faster. Discount for early payment if you have to. Every day you can extend the cash cycle helps.

Have a backup plan: A line of credit. A bridge loan. An investor who can write a check quickly. Don't wait until you're desperate. Set this up when you don't need it.

The Funding Decision Tree

When evaluating funding options, use this decision tree (on the next page).

The tree isn't rigid. Your situation might span multiple paths. But it forces you to think through your actual constraints and options instead of defaulting to "we need to raise money."

The Funding Mistakes That Kill Scaling Companies

Mistake 1: Raising too early

You raise money before you understand your model. You scale on assumptions. The model doesn't work. You run out of money before you can fix it.

Wait until you have proof. Then raise to scale what works.

Mistake 2: Raising too much

More money feels safer. It's not. It enables waste. You hire too fast. You spend carelessly. You lose discipline.

Raise what you need plus a buffer. Not more.

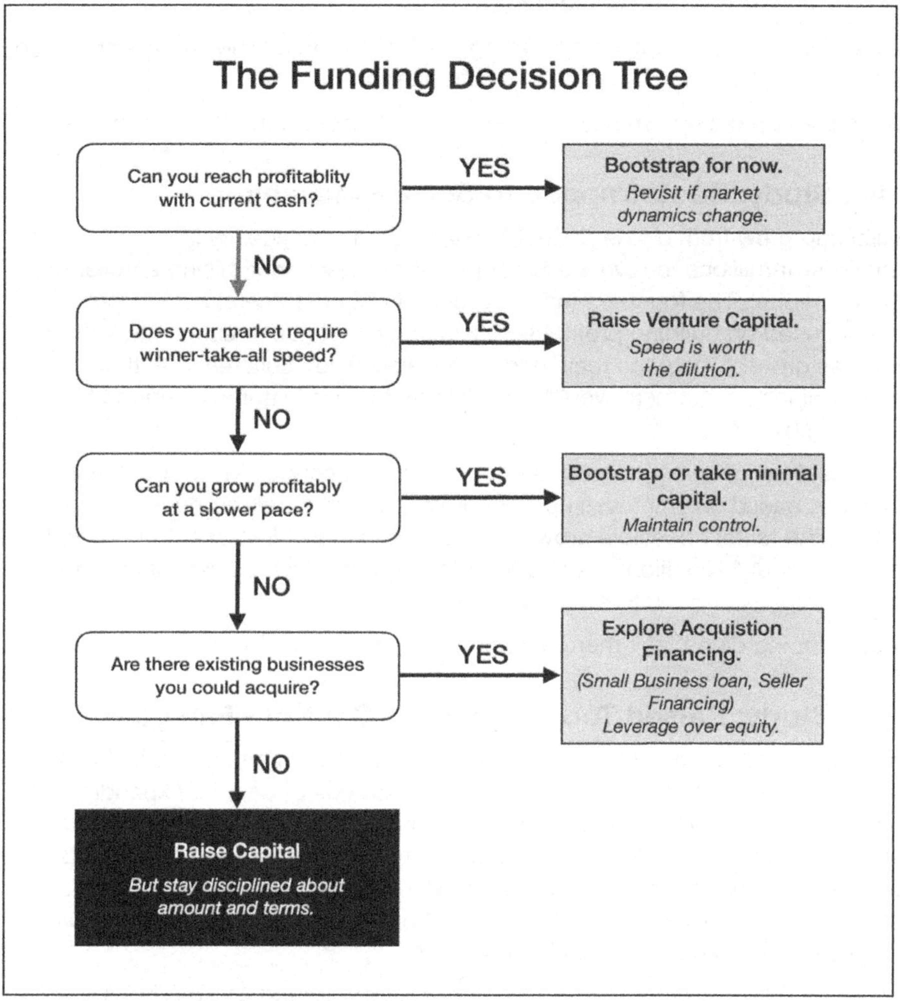

Mistake 3: Taking money from the wrong people

Investors who don't understand your business will push you in wrong directions. Investors with short timelines will pressure you to exit before you're ready. Investors who can't help beyond cash are just expensive.

Choose investors for strategic value, not just capital.

Mistake 4: Ignoring cash flow

You watch revenue and profit. You ignore cash. You run out of money while showing growth.

Cash is what matters. Model it. Track it. Protect it.

Mistake 5: Not having a backup plan

You assume everything goes according to plan. It won't. When things break, you scramble for emergency funding on terrible terms.

Set up backup capital when you don't need it. Hope you never have to use it.

Case Study: Bootstrapped to Scale - Mailchimp

Mailchimp grew from a side project inside a small web agency into a hundreds-of-millions-in-revenue SaaS business, all without raising a dollar of venture capital. The founders funded growth from customers, not investors. They focused on building profitable, sustainable revenue streams and let the business grow at the pace their income allowed. They built features their small-business customers would actually pay for, not features designed to impress a board.

Because they never sold equity, the founders kept control. They could make decisions based on what was right for their users and their product, instead of chasing the fastest possible growth curve. When Mailchimp eventually sold to Intuit for about $12 billion, the two founders together owned essentially the entire company and captured most of that value.

Bootstrapping didn't limit them. It focused them.

Case Study: Raised Too Much, Lost Control - Fab.com

Fab raised more than $300 million at a peak valuation of around $1 billion. With that cash, they scaled aggressively—hiring hundreds of people, expanding into new markets, and piling into inventory and marketing before they had a stable model. The funding made it easy to say yes to big bets and constant pivots instead of forcing discipline. Strategy shifted repeatedly as Fab tried to find something that worked, and the company burned through most of the capital without ever building a sustainable business.

In the end, Fab's ecommerce assets were sold for around $15 million, a tiny fraction of its former valuation. Investors lost hundreds of millions, and the founders had long since given up control.

More money didn't fix an unclear strategy; it just let them make bigger mistakes, faster, and at scale.

Case Study: Smart Acquisition Financing - Walker Deibel

Walker Deibel built a portfolio of businesses through acquisition entrepreneurship. Instead of starting from scratch, he used a combination of SBA loans and seller financing to buy existing, profitable companies with relatively little cash out of pocket. Each acquisition came with established revenue, customers, and cash flow that could help service the acquisition debt.

He focused on improving and growing the companies he bought—tightening operations, expanding sales, and building value—without raising venture capital or giving up control to outside investors. His primary cost of capital was debt service, not ownership dilution. That structure allowed him to retain equity, stay in control of his companies, and build wealth much faster than if he'd tried to bootstrap comparable businesses from zero.

My Story: The Funding Miscalculations

We started bootstrapped. Our first four agents generated enough revenue to sustain operations. We were profitable, not rapidly growing, but sustainable.

When we decided to scale to 60 agents, we faced a funding question. We could:

1. Raise capital to fund the growth
2. Scale slowly on revenue
3. Try to bootstrap rapid growth

We chose option three. Bad choice.

We didn't model the cash implications properly. We calculated that 60 agents would generate 15x the revenue of 4 agents. Simple math.

What we didn't model:

Ramp time: New agents weren't productive immediately. They took months to become effective. During that time, they were pure cost. We paid salaries for 2-3 months before they generated meaningful revenue.

Underperformers: We assumed everyone would perform at average levels. Many didn't. Some never got close. We paid their salaries for months before we faced reality.

Training costs: Training 56 people took time and resources. We pulled our best agents off the phones to help train. That reduced revenue from our highest performers.

Fixed costs: More people meant more space, more equipment, more overhead. These costs hit immediately. Revenue came later, if at all.

By month three, we realized we were burning cash way faster than we projected. Our runway was shrinking fast. We had maybe 8 weeks of cash left.

We had to fire 36 people just to stabilize cash flow. That was traumatic for everyone. It hurt morale. It damaged our culture.

If we had modeled properly, we would have either:

1. Raised capital to fund the 60-person growth
2. Scaled to 15-20 people on revenue, showing it could work, then expanded further.

3. Looked for acquisition opportunities instead of building from scratch

Instead, we tried to bootstrap aggressive growth. We ran out of runway. We barely survived.

The lesson: know your funding needs before you scale. Don't assume you can bootstrap rapid growth unless you've modeled it conservatively. Get the capital you need, or grow at the pace your cash allows.

Moving Forward

Money is a tool. Used right, it accelerates growth. Used wrong, it enables failure at scale.

Decide what you're aiming to optimise. Control? Speed? Sustainability? Your choice shapes your funding route.

Plan your cash flow carefully before scaling. Be clear on the required runway. Create a buffer.

Choose investors wisely. Borrow from people who grasp what you're creating and can support you beyond just funding.

Have a backup plan. Set up access to capital before you need it.

The next chapter will show you how to use that capital to scale your marketing and sales. But capital without a repeatable growth model just funds expensive experiments.

Get the funding right. Then build the engine that uses it effectively.

6

Marketing And Sales At Scale

The sales approach that got you to your first million won't get you to ten million. The marketing that worked for your first hundred customers won't work for your next thousand.

This is one of the toughest transitions when scaling. Your early growth came from things that don't scale. Founder hustle. Personal relationships. Manual outreach. One-off deals. These work brilliantly at small scale. They become impossible at large scale.

Most founders know this intellectually. They struggle with it emotionally. The old way feels comfortable. It worked. Letting go means trusting that something new will work just as well. That's scary.

The companies that scale through this transition do two things: they build repeatable systems that replace founder magic, and they're willing to kill their darlings when the old way stops working.

The companies that fail keep doing what worked early, long past when it stopped working. They become bottlenecks on their own growth.

What Got You Here Won't Get You There

Your first customers came from hustle. You reached out personally. You built relationships. You customized everything. You closed deals through force of personality and deep product knowledge.

This is founder-led sales. It's powerful. It's also completely unscalable.

Our directory business started this way. The founder sold. Personal relationships with business owners. Custom pitches. High touch. It worked. We closed deals. We proved the model.

When we had four agents, we could still operate this way. Each agent could build relationships. They could customize. They could hustle. Our best agents were essentially mini-founders.

When we tried to scale to 60, this approach collapsed. We couldn't find 60 people with founder-level hustle. We couldn't train 60 people to build relationships the way we did. The model that worked at four couldn't work at 60.

We tried anyway. We hired people. We gave them phones. We told them to build relationships and sell. Most couldn't do it. The conversion rates plummeted.

The transition from founder-led to repeatable:

Document what you actually do: Record your sales calls. Write down your pitch. Map your process. Most founders "just know" how to sell their products. That knowledge lives in your head. Get it on paper.

Find the pattern: You're not magic. There's a structure to what you do. You probably ask similar questions. You handle objections similarly. You close similarly. Find that pattern. That's your repeatable process.

Test if others can execute it: Give your documented process to someone else. Can they get similar results? If not, you haven't documented the *real process.* You've documented the *surface*. Go deeper.

Simplify until it works: Your process at scale will be simpler than your founder process. That's okay. You're trading depth for repeatability. A simple process that works consistently beats a complex process that only works when you do it.

We never completed this transition. We documented our pitch. We created scripts. But we never simplified enough. Our process still required relationship-building skills that most people don't have. We needed a process that an average salesperson could execute. Instead, we kept trying to hire exceptional people.

Building Repeatable Acquisition Channels

At small scale, you can manually acquire every customer. At large scale, you need channels that bring customers to you or sales systems that work without your direct involvement.

The question isn't "what channels exist?" It's "which channels can we make profitable and repeatable for our business?"

Repeatable means:

You can increase spend and get proportional returns. If you spend $10K and get 50 customers, you can spend $100K and get 500 customers. The economics stay consistent.

You're not reliant on your personal relationships. The channel works because of the process, not because you know people.

New people can run it. You can hire someone, train them, and they can execute the channel without constant oversight.

The results are predictable. You know roughly what you'll get for what you spend. Variance exists but you're not gambling.

Different channels for different businesses:

Software companies might use: content marketing, paid search, product-led growth, partnerships, and inside sales teams.

Manufacturing companies might use: trade shows, direct sales teams, distribution partnerships, and industry publications.

Service businesses might use: referral programs, speaking engagements, content marketing, outbound sales, and strategic partnerships.

Retail businesses might use: local advertising, social media, events, word of mouth, and loyalty programs.

The channel matters less than whether you can make it repeatable and profitable. One great channel beats five mediocre ones.

How to test channels:

Pick one at a time: Don't test everything simultaneously. You won't learn what works. Focus on one channel. Master it or kill it. Then move to the next.

Set a clear test: Define what success looks like. "We'll spend $5K on LinkedIn ads. We need to acquire 25 customers at $200 CAC or less. If we hit that, we'll scale it. If not, we'll try something else."

Give it enough time: Some channels work immediately. Others take months. Content marketing needs time to compound. Paid search can work in weeks. Know the channel's typical timeline before you judge it.

Track everything: You need to know CAC by channel. LTV by channel. Payback period by channel. Conversion rate at each stage. Without data, you're guessing.

Scale what works, kill what doesn't: Once you find a channel that hits your targets, double down. Hire someone to own it. Build systems around it. Run it until it stops working. Kill channels that don't hit targets after a fair test.

Our outbound cold-calling model worked initially. But it wasn't repeatable at the scale we aimed for. We should have tested whether the model could succeed with average performers before hiring 60 people. We didn't. We assumed it would scale in the same way it did with high-level performers at a smaller scale.

The Transition from Founder-Led Sales

This is where most B2B companies struggle. The founder closes all the big deals. Customers want to talk to the founder. The founder's presence signals importance.

You can't scale this. Eventually, the founder becomes the bottleneck. Deals stall waiting for the founder's time. The sales team depends on the founder to close.

How to transition:

Bring salespeople into your deals: Don't close alone. Bring a sales rep. Let them watch. Let them participate. Eventually, let them lead while you support. Then let them close without you.

Create deal structures that don't require you: If customers insist on meeting the founder, something's wrong. Either you're selling on personal relationships instead of value, or you haven't positioned your team properly. Fix this before you scale.

Promote your team's expertise: In calls and meetings, highlight your team's knowledge. "Sarah is our expert on implementation. She'll walk you through how this works." Transfer authority from yourself to your team.

Step back publicly: Tell customers you're stepping back from day-to-day sales. "I'm moving to product strategy. Sarah will be your main contact. She's better at this than I am." Make it normal.

Accept that some deals will be lost: When you step back, you'll lose some deals you would have closed. That's okay. You're trading short-term revenue for long-term scalability.

For service businesses, this might mean productizing your offering so it doesn't depend on founder expertise. For product businesses, it means building a sales process that works without founder credibility.

Brand Evolution: When to Keep Your Positioning, When to Change It

Your early brand was probably scrappy, personal, and founder-driven. At scale, you need something different.

What to keep:

Your core value proposition. If customers choose you for quality, maintain quality. If they choose you for speed, maintain speed. The reason customers buy shouldn't change.

Your authentic voice. Don't suddenly sound corporate. If you started casual and direct, stay casual and direct. Your early customers loved that.

What to change:

The emphasis on your founder. Early marketing is often "meet our founder." Later marketing is "meet our solution." Shift the focus from person to product.

Your messaging sophistication. Early messaging is often feature-focused. "Here's what we do." Later messaging is benefit-focused. "Here's what you'll achieve."

Your targeting breadth. Early, you might target narrowly. One industry. One use case. As you scale, you might expand. Or you might stay narrow and go deeper. Either way, your positioning evolves to match your strategy.

Industry considerations:

Tech companies often need to evolve from "scrappy startup" to "enterprise-ready." Enterprise customers want stability signals.

Manufacturing companies evolve from "custom fabrication" to "reliable partner at scale."

Service businesses evolve from "expert founder" to "expert team."

The evolution should feel natural, not forced. You're not becoming someone else. You're becoming a more professional version of who you were.

Inheriting Customers and Channels Through Acquisition

When you buy a business, you inherit their customers and their acquisition channels. This can accelerate growth dramatically or create integration nightmares.

What you get:

Existing customers who are already paying. This is immediate revenue and validation that the model works.

Acquisition channels that are already functioning. They've figured out how to reach customers. You can learn from and expand their approach.

Brand presence in their market. They have reputation, relationships, and trust. You can build on this.

What to watch for:

Customer concentration risk. If 80% of revenue comes from three customers, you don't have a stable base. You have fragility.

Founder-dependent relationships. If the previous owner sold everything through personal relationships, those customers might not stay when they leave.

Dying channels. The channels that worked for them might be declining. They might have sold because they couldn't scale. Don't assume their channels will work for you.

Integration approach:

Keep what's working. If their acquisition channel is more effective than yours, learn from it. Consider expanding it.

Merge carefully. Don't immediately force your brand and approach on their customers. They bought from the business you acquired, not from you. Transition gradually.

Find synergies. Can you cross-sell your products to their customers? Can their channels reach your target market? Look for ways to combine strengths.

The Scaling Marketing Playbook

The transition from founder-led to systematic marketing and sales happens in predictable phases. Understanding where you are and what comes next helps you build the right systems at the right time.

Most founders try to skip phases or cling to approaches that worked earlier. The result is chaos. Either they're building enterprise systems when they have 20 customers, or they're still doing everything manually at 500 customers.

This playbook shows the natural evolution. Use it to identify where you are now and what needs to change before you reach your next milestone.

The Scaling Marketing Playbook

Phase 1: Founder-Led 0-50 Customers	Phase 2: Process Emerging 50-250 Customers	Phase 3: Repeatable System 250-1,000 Customers	Phase 4: Multi-Channel Scale 1,000+ Customers
Primary Channel: Founder and direct outreach	**Primary Channel:** Founder and early sales team	**Primary Channel:** 2-3 proven channels at scale	**Primary Channel:** Portfolio of channels
Sales Process: Custom for each deal	**Sales Process:** Documented but customized	**Sales Process:** Standardized, clear stages	**Sales Process:** Different processes per segment
Goal: Prove people will pay	**Goal:** Repeat what worked	**Goal:** Profitable acquisition at volume	**Goal:** Efficient growth across markets
Success Metric: Any revenue	**Success Metric:** Consistent conversion rates	**Success Metric:** CAC and LTV by channel	**Success Metric:** Blended CAC, market share

Each phase requires different skills and systems. Don't try to skip phases. Master one before moving to the next.

Case Study: Cracking Repeatable Growth - HubSpot

HubSpot's early years followed a typical founder-led-sales pattern: the founders and small team were deeply involved in selling, guiding early customers, and shaping the product. Over time, they realized that this model didn't scale if every deal depended on the founders' time and instincts.

They built a repeatable inbound marketing engine—SEO, blog content, and free tools that generated leads consistently. Then they created an inside sales team that could convert those leads systematically, using documented playbooks for sales scripts, objection handling, and demo flows. As they grew, they hired people and trained them to plug into that system, turning founder-led chaos into a repeatable, teachable process.

The result was a company that scaled from an early-stage startup to billions in revenue with a model that didn't depend on founder magic. Instead, it depended on a repeatable, documented system that turned marketing and sales into an engine.

Case Study: Stuck in Founder Mode - Many Consulting Firms

Professional services firms often struggle with this transition. The founder is the expert. Clients want the founder. The founder bills most of the hours.

The business grows to the limit of the founder's capacity. Then it stops. The founder can't sell more than they can personally deliver.

Firms that break through productize their expertise. They train others to deliver. They build a brand around the firm, not the founder. They create frameworks that systematize their approach.

Firms that don't break through stay boutique. There's nothing wrong with that. But it's not scaling. It's hitting a natural ceiling and staying there.

Moving Forward

Scaling marketing and sales means building systems that work without you. It means finding channels that are repeatable and profitable. It means transitioning from founder-led to team-driven.

This is hard. It feels like you're losing control. You're giving up the approach that got you here. You're trusting systems over hustle.

But you can't scale yourself. You can only scale systems.

Document your approach. Test channels systematically. Build a sales process that average performers can execute. Transition from personal selling to systematic selling.

The next chapter will cover the operational side of scaling. But all the operational efforts in the world won't help if you can't acquire customers profitably and repeatedly.

Get your acquisition engine working. Then scale the operations to support it.

7

Operations and Execution

Running Faster Without Falling Apart

You can have perfect strategy, solid metrics, great people, and enough capital. None of it matters if you can't execute.

Operations is where strategy meets reality. It's where plans become actions. It's where you deliver what you promised to customers. Get operations right and you can scale smoothly. Get it wrong and everything falls apart.

Most founders underestimate operations. It feels less exciting than product development or sales. It's not the vision. It's just getting stuff done.

But operations are what separate companies that scale successfully from those that implode under their own growth. The sexy stuff attracts customers. Operations keep them.

We learned this the hard way. We had sales scripts. We had a training program. We had a plan. What we didn't have was the operational capability to execute that plan with 60 people instead of four.

We couldn't track individual performance effectively. We couldn't coordinate that many people. We couldn't maintain quality standards. Our operations collapsed under the weight of rapid growth.

Process vs. Agility: The Scaling Paradox

Small companies move fast. No approvals needed. No bureaucracy. See a problem, fix it. See an opportunity, grab it. This agility is your competitive advantage.

Then you scale up. You add people. You add complexity. Decisions that one person made now need coordination across five people. What once took an hour now takes a week.

You need process to coordinate. But process can slow you down. Too much process makes you a slow-moving bureaucracy. Too little and you end up with chaos.

The companies that scale well find the balance. They build processes that enable speed instead of preventing it.

How to stay nimble while growing:

Process the repeatable, not the unique: Standard operations need standard processes. Customer onboarding should follow a process. Quality checks should follow a process. These happen over and over. Process makes them consistent and efficient.

Creative work and strategic decisions shouldn't follow rigid processes. Product development needs experimentation. Strategic pivots demand quick decisions. Don't impose a process where flexibility is more important.

Develop decision-making frameworks, not approval chains: Instead of saying "everything needs three approvals," establish frameworks such as "You can spend up to $5K without approval if it solves a customer problem." People can move fast within clear boundaries.

Default to action: In ambiguous situations, bias toward doing something instead of waiting for perfect clarity. You can correct mistakes. You can't recover lost time.

Keep teams small: Amazon's two-pizza rule works. If a team can't be fed by two pizzas, it's too big. Small teams coordinate naturally. Large teams need meetings and processes.

Review and simplify regularly: Every quarter, look at your processes. Which ones help? Which ones exist just because they always have? Kill the useless ones.

Industry affects this differently. Software companies can maintain more agility because changes deploy quickly. Manufacturing companies need more processes because physical operations require coordination. Service businesses fall in between, depending on how standardized their service is.

Quality Control When You Can't Touch Everything

At four people, quality control is simple. You see everything. You touch everything. Problems are obvious. You fix them immediately.

At 60 people, you can't see everything. Work happens that you never touch. Quality becomes a system problem, not a personal attention problem.

This transition breaks many founders. They try to maintain hands-on quality control at scale. They become bottlenecks. They slow everything down. They burn out.

The solution isn't working harder to check everything. The solution is building systems that ensure quality without your direct involvement.

How to maintain quality at scale:

Define quality clearly: "High quality" is vague. What does it actually mean? What are the specific standards? Write them down. Make them measurable.

For a software company, quality might mean: fewer than 5 bugs per 1,000 lines of code, 99.9% uptime, response time under 200ms.

For a manufacturing company: defect rate under 0.5%, specifications within tolerance ranges, and on-time delivery above 95%.

For a service business: client satisfaction scores above 4.5/5, deliverables on time 90%+, budget variance under 10%.

For a sales operation like ours: daily sales rate of more than 1-sale per agent, customer retention above 95%, average sale value of $700.

Build quality into the process: Don't inspect quality at the end. Build it into each step.

Software companies rely on automated testing. Manufacturing employs in-process quality checks. Service businesses use peer review. Sales operations need performance dashboards that display problems in real time.

We didn't do this. We hired 60 agents and assumed all of them would perform well. We built a real-time system to track individual performance, but we ignored it and focused on the extra revenue coming in. By the time we realized quality had fallen, weeks had already gone by.

Create feedback loops: The people doing the work should see quality metrics. Not just managers. Everyone. When people see their own quality data, they self-correct.

Display individual dashboard performance. Show conversion rates, customer satisfaction, and defect rates. Make these metrics visible.

Empower people to stop production: If quality breaks, anyone should be able to stop the line. This is Toyota's andon cord principle. Better to stop and fix than to produce garbage at scale.

Spot check, don't inspect everything: You can't review every sales call or product unit. Choose samples randomly. Look for patterns. If spot checks show problems, investigate further.

Technology and Automation: What to Invest In (And When)

Technology should make operations easier. But most companies buy technology wrong. They buy too early or they buy the wrong tools.

Common technology mistakes:

Buying too early: You're a team of 10. You buy enterprise resource planning software designed for companies with 1,000 staff. It's overkill. It's costly. Nobody uses half the features.

Automating flawed processes: Your process is inefficient. You automate it. Now you have an inefficient process that runs faster. Fix the process first. Then automate.

Buying because competitors have it: "Everyone uses Salesforce." Maybe. But do you need what Salesforce does? Or would a simple spreadsheet work fine for another year?

Letting vendors define your workflow: You implement their software, and their workflow becomes your workflow, even if it doesn't match how you actually work.

Smart technology investment:

Start simple: Spreadsheets last longer than you think. Don't rush into buying software. When spreadsheets become difficult, then upgrade.

Buy for pain points: What's breaking right now? What's taking too much time? What's causing errors? Invest in technology that solves actual problems, not imaginary ones.

Buy for your current scale, not your planned future: You're 20 people now. You might be 200 in three years. Buy for 20. You'll upgrade before reaching 200 anyway.

Implement in stages: Don't turn on all features at once. Begin with the essentials. Ensure they work. Then introduce more complexity.

Measure the impact: Did the technology truly help? Is the problem resolved? Are people more productive? If not, you bought the wrong one.

What to automate first:

Repetitive data entry: If people copy information between systems, automate that. It's error-prone and soul-crushing.

Report generation: If someone spends hours building the same report every week, automate it. Time saved adds up.

Customer communications: Email sequences, appointment reminders, follow-ups. These can be templated and automated.

Monitoring and alerts: Set up systems that notify you when something fails. Don't rely on someone to spot it.

We needed better performance tracking technology. We relied on spreadsheets and reports, but lacked real-time dashboards that showed each agent's

Operational Scaling Checklist

Assess the maturity of your operational systems

1. Quality Systems

- ☐ Quality standards are documented and measurable
- ☐ Quality is checked during work, not just at the end
- ☐ Everyone can see quality metrics for their work
- ☐ People can stop work when quality breaks
- ☐ You spot-check samples regularly

2. Process Documentation

- ☐ Core processes are written down
- ☐ New people can learn processes from documentation
- ☐ Processes are reviewed and updated quarterly
- ☐ You know which processes are critical vs. nice-to-have

3. Technology & Tools

- ☐ Your tools solve actual pain points
- ☐ Technology enables work rather than creating overhead
- ☐ Data flows between systems without manual entry
- ☐ You have real-time visibility into operations
- ☐ Automation handles repetitive tasks

4. Coordination

- ☐ Everyone knows who owns what
- ☐ Work and progress are visible to the team
- ☐ You have structured communication rhythms
- ☐ Teams can work independently without constant checking
- ☐ Decisions are documented

5. Performance Tracking

- ☐ You track individual performance, not just team totals
- ☐ People get feedback on performance regularly
- ☐ Problems are caught early, not months later
- ☐ You know your capacity and constraints
- ☐ Bottlenecks are identified and addressed

performance and alerted us when it dropped. We should have invested in that before scaling to 60 people. Instead, we flew blind.

The Coordination Challenge: More People, More Problems

With four people, coordination comes easily. You chat. You work as a team. Everyone is aware of what the others are up to.

At 60 people, natural coordination breaks down. People don't know what others are working on. Work gets duplicated. Things fall through cracks. Miscommunication multiplies.

This is the coordination tax of scale. More people means more communication overhead. The question isn't whether coordination gets harder. It's how you manage it without drowning in meetings.

Coordination strategies:

Clear ownership: Every task, project, and decision should have a single owner. Not a committee. One person who is responsible for the outcome.

Transparent information: Use tools that show what's happening. Project management software. Shared dashboards. Status updates. Make information open instead of locked in people's heads.

Structured communication: Daily stand-ups for quick updates. Weekly team meetings for planning. Monthly all-hands for company direction. Don't let communication happen randomly.

Document decisions: When you decide something important, write it down. Where it was decided, why, and what happens next. People miss meetings. Documents persist.

Reduce dependencies: Design work so teams can operate independently. The more Team A needs Team B to finish before they can start, the slower everything moves.

Standard meeting practices: Prepare agendas before meetings. Take notes after meetings. Define clear action items with owners and deadlines. Most meetings are unproductive because they lack structure.

Different industries coordinate in different ways. Software firms rely on tools like Jira and Slack. Manufacturing uses production schedules and floor management. Service businesses depend on client management systems and project plans. The tools vary, but the core idea remains the same: make work visible and minimise friction.

The Operational Scaling Checklist

Operations readiness isn't about having everything perfect. It's having the foundations in place to support growth without breaking.

Too many founders scale operationally unprepared. They have great products and willing customers. But their operations can't handle the volume. Quality drops. Coordination fails. The whole thing falls apart.

Use this checklist (on the next page) before you scale to identify gaps in your operational readiness. You don't need to check every box. But if you're missing most items in any category, that's your weak point. That's what will break when you grow.

Be honest in your assessment. This isn't about looking good. It's about identifying problems before they turn into crises.

If you can't check most of these boxes, you're not ready to scale operationally. Build these foundations first.

Integrating Operations Post-Acquisition

Buying a business means inheriting its operations. Sometimes these operations are top-notch. Other times, they're just held together with duct tape and the founder's heroics.

What to assess:

Process maturity: Are operations well documented? Can they continue functioning without the current owner? Or is all knowledge confined to one person's head?

Technology stack: What systems do they use? Are they modern and maintainable, or legacy systems that will need replacement?

Team capability: Can the current team handle operations, or were they relying on the owner to do everything?

Hidden dependencies: What breaks when the owner leaves? What relationships or knowledge can't be handed over?

Integration approaches:

Keep what works: If their operations are better than yours, learn from them. Don't force your methods simply because you bought them.

Fix what's broken: If operations are weak, bring your systems in. But do it thoughtfully. Don't disrupt everything at once.

Find the middle ground: Perhaps they excel at some things. You excel at others. Combine the best of both.

Move slowly: Operations changes affect customers. Don't rush. Test changes gradually before full rollout.

The opportunity in acquisition often lies in operational improvement. You buy a business with quality products and loyal customers but weak operations. You bring operational discipline. You unlock value. This approach works if you're honest about what is broken and patient in fixing it.

Case Study: Operational Excellence at Scale - Amazon

Amazon scaled from an online bookstore to a vast, global 'everything store' while maintaining a remarkably high level of operational excellence. How?

They obsessed over operations, automated relentlessly, and built systems capable of handling massive volume across logistics, warehousing, and technology. They measured key metrics continuously and used data to optimize every major process. Clear ownership is baked into the culture: almost every major initiative has an owner, and every important metric has accountability, so problems cannot be ignored.

Even at enormous size, Amazon stays nimble through small, autonomous teams with clear decision-making authority. These teams run experiments, launch features, and iterate quickly, which keeps the organization moving fast while the underlying systems absorb the scale.

The result is a company that can coordinate millions of products, thousands of suppliers, and billions of customer interactions every year without collapsing under its own complexity—a rare example of operational excellence at true scale.

Case Study: Operations Collapse Under Growth - Theranos

Theranos expanded its promises far faster than its operations could actually deliver. The company claimed its technology could run hundreds of blood tests quickly and accurately from a tiny finger-prick sample, but the underlying devices and lab processes did not live up to those claims. Quality control and testing protocols were weak, and the organization struggled to coordinate and maintain consistency across its labs and infrastructure.

Instead of fixing these operational failures, Theranos concealed them. It showed investors and partners what they wanted to see—impressive demos and glossy promises—while hiding the flawed technology and unreliable results. The gap between the narrative and reality kept widening.

The operations simply could not support the growth and scale Theranos advertised. When regulators and journalists began to uncover the truth, the company collapsed, patient results were invalidated, and its founders were convicted of fraud and sentenced to prison.

Theranos is a brutal reminder: in high-stakes domains, operational capability isn't optional. You can't pretend to have it; the system either works, safely and reliably, or it fails, often catastrophically.

Moving Forward

Operations isn't glamorous. It's not the vision or the strategy or the pitch. But it's what determines whether you can actually deliver what you promised.

Build quality into your process. Invest in technology that addresses real issues. Develop coordination systems that work at scale. Monitor individual performance, not just overall figures.

Getting operations right makes scaling easier. Getting it wrong turns growth into a nightmare.

The next chapter addresses what happens when operations and everything else isn't working. When the metrics tell you scaling isn't going as planned. When you need to make hard decisions about whether to push forward or pull back.

Build strong operations now. You'll need that foundation when things get difficult.

8

Reading The Warning Signs

When Scaling Isn't Working

This is the chapter most scaling books don't write. The one about failure.

Everyone wants to talk about success. How to scale from 10 to 100 people. How to go from $1M to $10M in revenue. How to build the next unicorn.

Nobody wants to talk about what happens when it's not working. When the metrics are headed in the wrong direction. When you're running out of cash and options. When scaling is failing and you need to make hard decisions.

But this is the most important chapter for most founders. Because many scaling attempts don't succeed on the first go, or even the second. Most founders face a moment where they must decide: keep pushing, pull back, change direction, or shut down.

That moment is terrifying. You've invested time, money, reputation. You've built a team that depends on you. You've made promises to investors, customers, and partners. Admitting it's not working feels like admitting you failed.

Here's the truth: recognizing when scaling isn't working and making a tough decision isn't failure. It's leadership. Driving your company off a cliff because you refused to read the warning signs? That's failure.

The companies that survive and ultimately succeed are the ones whose founders can face reality and take action. Those that fail are the ones whose founders kept hoping until there was nothing left.

We faced this moment. Our metrics told us scaling wasn't working. We ignored those signals for too long. By the time we acted, we had to fire 36 people just to survive. If we'd read the warning signs earlier and acted decisively, we could have avoided that trauma.

The Metrics That Scream "STOP"

In Chapter 2, we talked about the metrics that matter when scaling is working. Now let's talk about the metrics that tell you it's not.

These aren't minor problems. These are red flags that mean you need to stop, assess, and make hard decisions immediately.

Metric 1: Deteriorating Unit Economics

Your CAC is climbing. Your LTV is dropping. Your LTV:CAC ratio is moving toward 1:1 or worse.

This means you're losing money on every customer. Growth is making the problem worse, not better. Every new customer digs the hole deeper.

What it tells you: Your model is broken. You're either targeting the wrong customers, charging too little, serving too inefficiently, or spending too much to acquire.

What to do: Stop scaling immediately. Figure out which part of the equation is broken. Fix it before expanding further.

We watched our cost per sale climb as we added agents. We saw our average sale value stay flat. Our cost-to-revenue ratio got worse every month. These metrics were screaming at us. We ignored them because we were focused on total revenue.

Metric 2: Accelerating Cash Burn with No Path to Profitability

Your monthly expenditure is growing faster than your income. Your runway is shrinking. You can't see a practical route to profitability at your current trajectory.

What it tells you: You're running out of time. You either need more capital, lower burn, or both. And "more capital" only helps if it extends runway to profitability. If you can't get profitable even with more money, more money just delays the inevitable.

What to do: Explore different scenarios. What burn rate allows you to reach profitability? What would you need to cut to achieve that? Can you still execute your plan at that burn rate? If not, you need a different plan.

We burned through our runway without realizing how fast it was going. By the time we looked seriously at the numbers, we had eight weeks left.

Metric 3: Declining Quality or Customer Satisfaction

Your customer complaints are increasing. Your Net Promoter Score is dropping. Your retention is decreasing. Customers who used to love you are leaving or raising complaints.

What it tells you: You're growing too quickly for your ability to maintain quality. You're breaking what made you special.

What to do: Stop growing. Fix quality. You can't build a business on unhappy customers. If you're losing customers as fast as you're gaining them, growth is pointless.

Metric 4: High Churn or Failure to Repeat

For subscription businesses: churn rate above 5% monthly. For transaction businesses: one-time buyers with no repeat purchases. For service businesses: low referral rates and declining renewals.

What it tells you: Customers aren't getting value. Your product or service isn't addressing their problem well enough to keep them loyal.

What to do: Stop selling. Fix the product or service. Talk to churning customers. Find out why they're leaving. Fix those problems. Then sell again.

Metric 5: Revenue Per Employee Declining

You're adding people faster than you're adding revenue. Your revenue per employee is dropping.

What it tells you: You're not scaling. You're just growing headcount. You're getting less efficient, not more efficient.

What to do: Stop hiring. Figure out why productivity is dropping. Are you hiring the wrong people? Training poorly? Managing badly? Fix the productivity problem before you add more people.

This was our killer metric. We went from four productive agents to 60 mixed-productivity agents. Our revenue per agent plummeted. We didn't watch this number. If we had, we would have stopped hiring after agent 15 or 20.

Metric 6: Founder Working Harder, Results Getting Worse

You're working 80-hour weeks. You're in every meeting. You're fighting every fire. But the metrics keep getting worse.

What it tells you: Working harder won't fix this. The problem is structural, not about effort. You're trying to solve a system problem with personal heroics.

What to do: Step back. The problem is the model, the strategy, or the execution approach. More of what's not working won't make it work.

The Honest Conversation: Pivot, Pause, or Pull the Plug?

When the metrics tell you scaling isn't working, you have four options:

Option 1: Pivot

Change something fundamental about your approach. Different target market. Different pricing. Different product. Different sales motion. Different everything.

When to pivot:

- The underlying opportunity is still valid
- You have runway to execute the pivot
- You understand what's not working and what might work instead
- Your team can execute the new direction

When not to pivot:

- You're out of runway
- You're pivoting because you're desperate, not because you have a clear hypothesis
- You've pivoted three times already
- You don't know what to pivot to

Option 2: Pause

Stop scaling. Go back to what works. Stabilize. Figure out what broke. Fix it. Then try again.

When to pause:

- You have enough cash to survive at smaller scale
- You know what broke and how to fix it
- Your team can survive the pullback
- Customers will stay with you at smaller scale

When not to pause:

- Pausing means running out of money
- The market window is closing and competitors are winning
- Your team will quit if you pause
- You don't know how to fix what's broken

Option 3: Sell

Find an acquirer. Sell the business. Exit before you crash.

When to sell:

- You have something valuable even if scaling isn't working
- Acquirers exist who can scale it better than you
- Your team and customers would be better off with new ownership
- You're emotionally ready to let go

When not to sell:

- The business is too broken to have value
- You'd be selling at fire-sale prices out of desperation
- You still believe you can fix it and have runway to try
- Selling would leave customers and team in a bad position

Option 4: Shut Down

Close the business. Return remaining capital to investors if applicable. Move on.

When to shut down:

- No path forward exists
- Continuing would just lose more money
- No acquirer would buy even at low prices
- You're out of runway and options

When not to shut down:

- You have other viable options
- You're reacting emotionally, not rationally
- You haven't tried other options yet
- Shutting down would be premature

The hardest part isn't knowing these options exist. It's being honest about which one applies to you.

The Pivot Decision Matrix

When scaling is failing, the hardest part isn't knowing what your options are. It's knowing which option is right for your specific situation.

Most founders make these decisions emotionally. They're attached to their plan. They're scared of admitting failure. They're influenced by others' expectations. They choose based on hope instead of reality.

This matrix removes emotion from the decision. It forces you to answer two simple questions honestly: How much time do you have? How confident are you that you know what's broken and how to fix it?

Your answers tell you what to do next. Not what you wish you could do. What you actually should do given your constraints.

Use this when the metrics from earlier in this chapter are telling you something's wrong. Don't wait until you're desperate. The earlier you use this framework, the more options you'll have.

The Pivot Decision Matrix

Diagnosis Confidence

Runway Remaining	High Confidence (*Clear Diagnosis / Fix*)	Medium Confidence (*Clear Diagnosis / Unclear Fix*)	Low Confidence (*Unclear Diagnosis*)
12+ Months (*Time to fix*)	PAUSE & FIX • You have time to fix it properly • Stop scaling • Fix what's broken, restart	PIVOT • Test new approaches • You can afford to experiment	SEEK HELP / SELL • You don't know what's wrong • Get outside perspective • Or find acquirer to fix it
12 Months (*Act quickly*)	FIX FAST • You know what to do • Execute quickly	HARD PIVOT • Make one big bet • If it fails, prepare to sell or shut down	SELL NOW • Don't wait until you have no value left • Maximize remaining asset
< 6 Months (*Emergency*)	EMERGENCY ACTION *Confidence level is irrelevant. Cash is the only priority.* • Massive cost cuts • Seek immediate capital • Start sale process • Prepare for shutdown		

Be brutally honest with yourself. Lying about your runway or your confidence doesn't hurt the matrix. It hurts you.

The matrix isn't rigid. Your situation might span multiple boxes. But it forces you to face two hard truths: how much time you actually have, and whether you actually know what to do with it. Most founders overestimate both. Don't be like most founders.

Scaling Failure Patterns (And How to Recognize Yours)

Most scaling failures follow patterns. Recognizing your pattern helps you know what to fix.

Pattern 1: Premature Scaling

You scaled before proving the model. You built for 100 customers when you had 10. You hired before you had repeatable revenue.

Signs: High burn, low revenue, young business, everything feels chaotic.

Fix: Pull back. Prove the model at small scale. Then scale.

Pattern 2: Scaling Broken Unit Economics

You scaled a model that loses money on each customer. Growth made it worse.

Signs: Revenue growing but losses accelerating, LTV:CAC below 2:1, gross margins too thin.

Fix: Stop growing. Fix unit economics. Then scale.

This was our pattern. We scaled while per-agent economics were deteriorating. More agents meant more losses.

Pattern 3: Scaling Without Systems

You tried to scale without building operational capability. Quality collapsed. Coordination failed.

Signs: Customer complaints rising, team overwhelmed, founder doing everything, chaos.

Fix: Pause growth. Build systems. Then scale.

Pattern 4: Scaling Into the Wrong Market

You're targeting customers who don't value what you offer or who cost too much to serve.

Signs: High CAC, low retention, customer complaints about fit, struggling to explain value.

Fix: Pivot to different customers or reposition offering.

Pattern 5: Scaling with the Wrong Team

You hired people who can't execute at the level you need. Performance is consistently below standard.

Signs: Revenue per employee dropping, quality issues, low productivity, constant management.

Fix: Upgrade team or simplify model to match team capability.

This was part of our problem as well. We hired people who couldn't perform at our standards. We kept them hoping they'd improve. They didn't.

Recovery Strategies: Scaling Back to Move Forward

Sometimes the path forward requires going backward first. Scaling back isn't failure. It's recalibration.

Strategy 1: The Right-Sizing

Cut back to the size where you're profitable and operational. Stabilize. Then grow again from solid foundation.

How:

- Model profitability at different sizes
- Identify which people and costs are essential
- Make cuts decisively, not gradually
- Communicate clearly why you're doing it

- Rebuild from stability

This is what we did. We cut from 60 agents to 24. We got back to profitability. We could breathe again. We could think. We could rebuild properly.

Strategy 2: The Focus Play

You tried to do too much. Cut everything except your core. Do one thing excellently instead of five things poorly.

How:

- Identify your one most valuable offering
- Cut other products/services/markets
- Reallocate resources to the core
- Become excellent at that one thing
- Expand again later from strength

Strategy 3: The Margin Recovery

You competed on price and killed your margins. Raise prices. Accept losing some customers. Keep profitable ones.

How:

- Analyze customer profitability
- Identify who pays fair prices and who doesn't
- Raise prices to sustainable levels
- Accept that some customers leave
- Focus on serving profitable customers excellently

Strategy 4: The Process Rebuild

You scaled sloppy processes. Pull back. Document and fix everything. Scale the fixed process.

How:

- Reduce volume to manageable level
- Document every process properly
- Train people to the process
- Verify quality is consistent
- Then scale the working process

Strategy 5: The Acquisition Exit

Sometimes the right answer is selling to someone who can scale it better than you.

When this works:

- You've built something valuable
- You're not the right person to scale it
- An acquirer can unlock value you can't
- Your team and customers would be better off

How:

- Start the sale process before you're desperate
- Be honest about what's broken
- Focus on the value you've created, not the potential you haven't achieved
- Negotiate for team and customer continuity if you care about them
- Exit with dignity

The "When All Else Fails" Playbook

You've tried to scale. The metrics say it's not working. You need a practical plan for what to do next.

Week 1-2: Honest Assessment

Get clear on reality. No more hoping. No more wishful thinking.

Actions:

- Pull every metric. Look at trends, not snapshots.
- Calculate the runway at the current burn.
- Model what happens if nothing changes.
- List everything that's broken.
- Rank problems by impact.
- Assess whether you can fix them with current resources and runway.

Be brutally honest. Get an outside perspective. Ask someone you trust to review your numbers and tell you what they see.

Week 3-4: Decision Time

Based on your assessment, pick your path.

Actions:

- Share assessment with co-founders, board, or advisors.
- Discuss the four options: pivot, pause, sell, or shut down.
- Make a decision. Don't delay. Every week costs money you don't have.
- Commit to the path. Half-measures don't work.

If you choose Pause/Fix:

Month 1:

- Cut costs to extend runway. Fast. Deep cuts. Get to profitability or an 18-month runway minimum.
- Communicate clearly to the team what's happening and why.
- Identify the 2-3 problems you must fix.
- Build a plan to fix them with a timeline and success metrics.

Months 2-6:

- Execute fixes.
- Track metrics weekly.
- Adjust plan based on results.
- Prepare to scale again only when metrics prove you've fixed the problems.

If you choose Pivot:

Month 1:

- Define the pivot clearly. What's changing? What's staying the same?
- Model the new approach. What are the unit economics? What's the timeline to profitability?
- Assess team fit. Can the current team execute the pivot? If not, what changes are needed?
- Communicate with everyone affected.

Months 2-6:

- Execute pivot with MVP mindset. Test before you scale.
- Track new metrics relevant to the pivot.
- Give it a fair test, but set a decision point. "If we don't hit X by Y date, we'll try something else."

If you choose Sell:

Month 1:

- Clean up your books. Accurate financials matter.
- Document operations. Acquirers want to understand what they're buying.
- Identify potential acquirers. Who could benefit from owning your business?
- Prepare materials. Business overview, financials, customer data, team information.

Months 2-6:

- Reach out to potential acquirers. Stay professional.
- Run the business while selling. Don't let operations collapse during the sale process.
- Negotiate for what matters to you. Price, team continuity, customer care.
- Close the deal or move to the next option if the sale isn't happening.

If you choose Shut Down:

Month 1:

- Plan the shutdown. Legal obligations, customer transitions, team notifications, investor communications.
- Manage the message. Be honest and professional.
- Take care of your people. Help them find new roles if possible. Give good references.
- Handle customers properly. Give them time to transition. Return unused fees if applicable.

Month 2-3:

- Execute shutdown plan.
- Wrap up legal and financial obligations.
- Learn from the experience. What would you do differently? What did you learn?
- Move forward. This isn't the end of your career. It's the end of one chapter.

My Full Story: What We Should Have Done When the Metrics Turned Bad

Here's what actually happened and what we should have done instead.

Month 1: The Hiring Blitz Begins

What we did: We decided to scale aggressively. We launched a hiring blitz to bring in more sales agents. We proved the model with our first four agents. We saw the chance to boost revenue by adding more people doing the same work.

What we saw: New agents joining the team. Training sessions underway. Total sales beginning to tick up as we added more people. Activity everywhere. It felt like progress.

What we told ourselves: "This is what scaling looks like. More people, more sales, more growth."

What we should have done: Before we started hiring, we should have built an individual performance tracking system. A dashboard showing each agent's daily metrics: calls made, contacts reached, sales closed, and conversion rates.

We should have set clear performance benchmarks: you must hit X conversion rate within Y weeks, or you're not a fit for the role. We should have decided upfront that we'd only keep people who met those standards.

We had training and scripts. What we didn't have was real-time performance visibility and clear performance gates.

Month 2: Reaching 60 Agents

What we did: We kept hiring aggressively. By the end of Month 2, we had 60 agents in total. We'd gone from 4 to 60 in roughly two months. We were scaling rapidly.

What we saw: Total revenue was increasing. Not as quickly as our simple calculations suggested (if 4 agents generate X, then 60 should generate 15X), but it was still growing. The figure was rising. We took that as validation.

What we told ourselves: "Revenue is growing. We're on track. Some agents are still ramping up, but overall we're moving in the right direction."

What we weren't seeing: Individual agent performance varied wildly. Some agents were performing at the level of our original four. Many weren't even close. A few were averaging more than 2.5 sales per day, whereas others were selling at 0.5 or less.

We were watching total revenue. We weren't monitoring revenue per agent. We weren't measuring cost per sale. We weren't tracking which specific agents were producing results and which weren't.

What we should have done: Two weeks into having 30+ agents, we should have had data showing individual conversion rates. At that point, we would have seen that many agents weren't performing at acceptable levels.

We should have stopped hiring immediately. We should have concentrated on the agents we already had. We should have identified who was meeting standards and who wasn't. We should have coached those who could improve and let go of those who clearly weren't suited for the role.

Instead, we kept hiring to hit our goal of 60 agents. We were so fixated on the hiring target that we didn't track whether those hires were actually productive.

Month 3: Reality Hits

What we saw: By the middle of Month 3, we looked hard at our financials. Revenue was up, but not nearly enough. Our costs had scaled linearly - 60 salaries, 60 desks, all the overhead. But our revenue hadn't scaled at the same rate.

We were adding costs faster than we were adding revenue. We ran the numbers. At our current trajectory, we were going to run out of cash in a few months.

What we did: We finally built the individual performance analysis we should have had from day one. We tracked each agent's conversion rates, sales per day, and revenue generated versus cost.

The data was clear and brutal. About 60% of our team was underperforming. They weren't covering their own costs. We had roughly 24-26 agents who were actually productive at acceptable levels. The other 34-36 were costing us money every day they stayed on the team.

We made the decision. We started letting the underperformers go. It wasn't a single mass layoff. We did it over a few weeks during Month 3 with our Team Leaders involved in the decision-making process. We were honest about why. "You're not hitting the performance benchmarks we need. This isn't working out. We're going to part ways."

How the team responded: Our high performers understood. They'd seen who was performing and who wasn't. They knew we needed to make changes. Because we were open and honest about what was happening and why, morale stayed strong.

The good agents wanted the business to succeed. They were making good money. They saw that we were making hard decisions to ensure the business survived. They respected that. None of our strong performers left. They stayed committed.

What we should have done: We should never have let it get to Month 3 with that many underperformers on the team.

If we had tracked individual performance from week one, we would have known within the first two weeks of someone joining the team whether they could perform at our standards.

We should have made performance decisions weekly, not after two months. The agent shows low performance in week 1, so give them focused coaching in week 2. Still low performance in week 3? Part ways and wish them well. Don't keep them for three months hoping they'll improve.

If we had done that, we would have:

1. **Scaled to 20-25 top performers instead of 60 mixed performers.** We would have hired selectively, keeping only those who demonstrated the ability to execute.
2. **Avoided the cash crisis.** 25 productive agents would have generated strong revenue while keeping costs manageable. Instead, we had 60 agents generating weak revenue with unsustainable costs.

3. **Built sustainable operations.** We could have managed 25 people properly. We couldn't manage 60. We would have had better training, better coaching, better coordination.
4. **Made better individual decisions.** Letting someone go after two weeks of poor performance is a clear business decision. Letting 36 people go after three months feels like a crisis. Even though we handled it professionally, it was still traumatic.

The Core Lessons:

1. Monitor individual performance from day one. Don't wait until a crisis arises to set up performance dashboards. We should have had individual metrics visible daily before hiring agent #5.

2. Set clear benchmarks and stick to them. We should have said: "You must hit X conversion rate within two weeks or this role isn't a fit." Then we should have enforced that standard immediately.

3. Fast feedback is kinder than slow failure. Telling someone after two weeks that they're not right for the role gives them a chance to find something else. Keeping them for three months and then letting them go wastes both their time and yours.

4. Don't get attached to hiring targets. We wanted to hit 60 agents. That became the goal. We should have wanted to hit "as many productive agents as we can find and manage effectively." The number should follow performance, not the other way around.

5. Watch cost-to-revenue ratios, not just total revenue. Revenue was growing, so we thought we were succeeding. But our costs were growing faster. We should have watched the ratio from day one.

6. Build the tracking system before scaling up (and monitor it closely). We built our performance-tracking system early, before we started scaling. But we only looked at it closely in Month 3, when we were in trouble. We should have set it up in Month 0 so we couldn't have ignored warning signs as they developed. By the time we had a good, close look at the data, we had already made costly mistakes.

If I could go back, before we started hiring, I would have set clear standards for what "good performance" looks like. I would have committed to making weekly decisions about who stays and who goes based on data, not hope.

We would have scaled to 20-25 excellent agents over six months, rather than 60 mixed agents in two months. We would have built something sustainable instead of something that nearly killed us.

That's what reading the warning signs looks like. That's what acting on them means. I didn't do it early enough. I hope you will.

Case Study: Recognized Problems and Acted - Slack

Slack started as a gaming company called Tiny Speck, building a multiplayer game called Glitch. The game launched but never gained the audience it needed to be sustainable. The team was running out of steam and options, and Stewart Butterfield faced a clear choice: keep pushing on a failing product or try something else.

While developing the game, they had built an internal communication tool that quickly became indispensable for their distributed team. As the game faded, they realized the real value was in the tool, not the game. Butterfield and the team made the decision to pivot: Tiny Speck would become Slack, a messaging platform for workplace communication.

Butterfield recognized that the original idea was not working and acted decisively, redirecting the company toward the one thing that was already proving valuable. That pivot salvaged the business and eventually created billions in value, turning an internal tool into one of the most successful enterprise software stories of the 2010s.

Case Study: Ignored Warning Signs Until Collapse - MoviePass

MoviePass offered unlimited movie tickets for about $10 a month. The unit economics were catastrophic: the company paid theaters the full ticket price, so every customer effectively cost more than they paid in subscription fees. The math was obvious: the business model couldn't work at that price unless usage stayed very low, and usage instead exploded.

They knew the economics were broken. Analysts, investors, and internal projections all showed that MoviePass was losing money on every new user, yet the company chose to double down. It raised hundreds of millions of dollars and scaled aggressively, hoping that data, advertising, or partnerships with theaters would eventually make the numbers work.

They didn't fix the core problem. Instead, they gradually changed the terms: capping the unlimited plan, limiting access to popular movies, and restricting usage. Customers revolted, subscribers fled, and the company burned through its capital. The business effectively collapsed, and its parent company spiraled into financial and legal turmoil.

If MoviePass had honestly confronted its unit economics on day one, it would have had to choose: either raise the price to something sustainable or accept that the business model couldn't work at scale. Instead, it scaled a product it knew was losing money, and ended up paying the price.

Moving Forward

Reading warning signs isn't pessimism. It's realism. The companies that survive and thrive are led by founders who can face hard truths and make hard decisions.

Watch your metrics. All of them. Not just the ones that make you feel good.

Set clear decision points. "If we don't hit X by Y date, we'll do Z." Don't move the goalposts. Honor your decision points.

Get outside perspective. Advisors, board members, mentors who will tell you the truth. Don't surround yourself with cheerleaders.

Act early. The earlier you act on warning signs, the more options you have. Wait until you're desperate and you'll make desperate choices.

Scaling isn't a yes or no. It's not "succeed at scale or fail completely." Sometimes the best move is to scale back and rebuild. Sometimes it's to pivot. Sometimes it's to sell. And sometimes it's to shut down and try something else.

What matters is making the decision consciously, based on data and reality, not hope or attachment.

You've built something. That matters. What you do next matters too.

Make the hard call. Your future self will thank you.

Conclusion: Scale Smart, Not Just Fast

You made it to the end. That puts you ahead of most founders.

Most people don't prepare for scaling. They just jump into it. They hit some early success and believe growth will sort itself out. Then they wonder why everything starts to fall apart.

You're different. You read this book. You learned from my mistakes and the mistakes of others. You've got frameworks to work with. You've got questions to ask yourself. You've got warning signs to watch for.

Now you need to actually use what you've learned.

The Long Game Beats the Fast Burn

The startup world worships speed. Move fast and break things. Blitzscale. Grow or die.

Sometimes speed wins. If you're in a winner-takes-all market with massive network effects, you might need to scale fast before competitors do.

But most businesses aren't like that. Most businesses succeed by scaling smart by building solid foundations. By growing at a pace they can sustain. By making money instead of burning it.

Fast scaling that breaks your company isn't success. It's an expensive failure.

Smart scaling might look slower. But it's sustainable. You keep your culture. You maintain quality. You stay profitable. You build something that lasts.

I tried to scale fast. I hired 60 people in a few months. I didn't track the metrics that mattered. I hoped speed would solve problems that only discipline could fix.

It nearly destroyed us.

When we pulled back, got disciplined, and rebuilt properly, we built something sustainable. It took longer. It felt slower. But it actually worked.

Don't confuse activity with progress. Don't confuse growth with scaling. Real scaling makes you more efficient as you grow, not less.

Your Blueprint for What Comes Next

You know where you are now. Here's what to do next:

If you're not ready to scale yet:

Use Chapter 1's readiness scorecard. Be honest about where you are. Fix the gaps before you try to grow. Build your foundation properly. You'll scale faster later because you did the work now.

If you're ready to scale:

Start with metrics from Chapter 2. Build your dashboard. Know your unit economics thoroughly. Monitor individual performance, not just overall totals. These metrics will tell you if scaling is working before anything else will.

Build the systems from Chapter 3 before you need them. Document your processes. Train your people properly. Set clear performance standards. Maintain the discipline to enforce them.

Protect your culture from Chapter 4. Define your core values. Hire for them. Fire when people violate them. Everything else can evolve, but the core must stay constant.

Get your funding right from Chapter 5. Know what you need. Model your cash flow. Don't try to bootstrap rapid growth unless you've planned for it. Don't raise more than you need just because you can.

Build repeatable acquisition process from Chapter 6. Document what works. Test it with others. Simplify it. Make it independent of your presence. Then expand it.

Strengthen operations from Chapter 7. Track quality. Coordinate effectively. Invest in technology that addresses genuine issues. Develop systems that facilitate speed rather than hinder it.

Watch for warning signs from Chapter 8. Set decision points. Respect them. Act early when metrics tell you something's wrong. Have the courage to pause, pivot, or pull back when necessary.

The Questions Only You Can Answer

I can give you frameworks. I can share what worked and what didn't. I can show you the metrics to watch and the mistakes to avoid.

But I can't answer your specific questions. Only you can:

What does success look like for you? Is it building a billion-dollar company? Is it creating a sustainable business that provides good income and control? Is it building something to sell? Know what you're building toward.

What are you willing to sacrifice? Time with family? Personal income? Control? Peace of mind? Scaling costs something. Be clear about what you're willing to pay.

What matters more - speed or control? If competitors are winning the market, you might need to raise capital and move quickly. If you have time, you might bootstrap and keep control. Know what is important to you.

Are you the right person to grow this? Sometimes the founder who starts a company isn't the best person to take it to the next level. That's okay. You can hire a CEO, sell the business, or keep it small. Be honest about your strengths.

What will you do if scaling doesn't work? Have you considered your options? Do you have a plan B? Making decisions under pressure is more difficult than when you have time to think.

These aren't easy questions. But answering them before you scale will help you make better decisions as you scale.

Final Wisdom from the Trenches

I've succeeded at scaling and also experienced failure. Most of the mistakes in this book are personally mine. Here's what I know:

Metrics don't lie. Feelings do. When metrics and feelings clash, trust what the metrics show. I felt we were making headway, but the numbers said we were failing. Turns out, the metrics were right.

Act on problems early. Every problem I ignored got worse. Every hard decision I delayed became harder. The earlier you act, the more options you have.

Systems beat hustle. At small scale, you can hustle through anything. At scale, you need systems. Build them before you desperately need them.

Standards are more important than plans. We aimed to expand to 60 agents but didn't set standards for the team members. The absence of standards sabotaged our plan. Establish high standards and enforce them. The plan will take care of itself.

Your team watches what you do. You can claim you value performance, but if you keep poor performers, your team learns that performance doesn't matter. Your actions shape your culture, not your words.

Scaling shows what you don't know. You'll find gaps in your knowledge, your systems, and your team. That's normal. The question is whether you admit them and fix them, or ignore them and hope.

The tough calls are the important ones. The easy decisions don't hold much weight. The tough ones - firing someone, pulling back from growth, admitting you were wrong - they determine whether you succeed.

You can recover from scaling failures. If scaling doesn't succeed, it's not the end. We sacked 36 staff and barely made it. Then we rebuilt properly. Failure at scaling isn't permanent unless you give up.

Success is building something you're proud of. It's not hitting some arbitrary revenue goal or making someone else's list. It's about creating something that works, serves your customers, looks after your team, and makes you proud. Define what success means to you, not what others say it should be.

You've Got This

Scaling is hard. It will test you. You'll make mistakes. Things won't go as planned. You'll face times when you wonder if you can do this.

You can.

You've built something worth scaling. That's the hard part. Now you need discipline, honesty, and courage to do it properly.

Discipline to build systems and adhere to them. Honesty to recognise problems and admit them. Courage to make tough decisions when metrics demand it.

You've got the frameworks. You've got the knowledge. Now go put them to good use.

Think smart about scaling. Watch your metrics. Look after your people. Create something that endures.

The rest is up to you.

Made in the USA
Monee, IL
04 November 2026

67380410R00066